PROBABILISTIC REASONING IN MULTIAGENT SYSTEMS

This book investigates the opportunities in building intelligent decision support systems offered by multiagent distributed probabilistic reasoning.

Probabilistic reasoning with graphical models, also known as Bayesian networks or belief networks, has become an active field of research and practice in artificial intelligence, operations research, and statistics in the past two decades. The success of this technique in modeling intelligent decision support systems under the centralized and single-agent paradigm has been striking. In this book, the author extends graphical dependence models to the distributed and multiagent paradigm. He identifies the major technical challenges involved in such an endeavor and presents the results from a decade's research.

The framework developed in the book allows distributed representation of uncertain knowledge on a large and complex environment embedded in multiple cooperative agents and effective, exact, and distributed probabilistic inference.

Yang Xiang is Associate Professor of Computing and Information Science at the University of Guelph, Canada, where he directs the Intelligent Decision Support System Laboratory. He received his Ph.D. from the University of British Columbia and developed the Java-based toolkit WebWeavr, which has been distributed to registered users in more than 20 countries. He also serves as Principal Investigator in the Institute of Robotics and Intelligent Systems (IRIS), Canada.

PROBABILISTIC REASONING IN MULTIAGENT SYSTEMS

A Graphical Models Approach

YANG XIANG

University of Guelph

CAMBRIDGE
UNIVERSITY PRESS

PUBLISHED BY THE PRESS SYNDICATE OF THE UNIVERSITY OF CAMBRIDGE
The Pitt Building, Trumpington Street, Cambridge, United Kingdom

CAMBRIDGE UNIVERSITY PRESS
The Edinburgh Building, Cambridge CB2 2RU, UK
40 West 20th Street, New York, NY 10011-4211, USA
477 Williamstown Road, Port Melbourne, VIC 3207, Australia
Ruiz de Alarcón 13, 28014 Madrid, Spain
Dock House, The Waterfront, Cape Town 8001, South Africa

http://www.cambridge.org

First published 2002

Printed in the United Kingdom at the University Press, Cambridge

Typeface Times 11/14 pt.   *System* LaTeX 2_ε [TB]

A catalog record for this book is available from the British Library.

Library of Congress Cataloging in Publication Data
Xiang, Yang, 1954–
Probabilistic reasoning in multi-agent systems : a graphical models approach / Yang Xiang.
p. cm.
Includes bibliographical references and index.
ISBN 0-521-81308-5
1. Distributed artificial intelligence.   2. Bayesian statistical decision theory – Data processing.
3. Intelligent agents (Computer software)   I. Title.
Q337 .X53   2002
006.3 – dc21   2001052874

ISBN 0 521 81308 5 hardback

Contents

Preface

This book investigates opportunities for building intelligent decision support systems offered by multiagent, distributed probabilistic reasoning. Probabilistic reasoning with graphical models, known as Bayesian networks or belief networks, has become an active field of research and practice in artificial intelligence, operations research, and statistics in the last two decades. Inspired by the success of Bayesian networks and other graphical dependence models under the centralized and single-agent paradigm, this book extends them to representation formalisms under the distributed and multiagent paradigm. The major technical challenges to such an endeavor are identified and the results from a decade's research are presented. The framework developed allows distributed representation of uncertain knowledge on a large and complex environment embedded in multiple cooperative agents and effective, exact, and distributed probabilistic inference.

Under the single-agent paradigm, many exact or approximate methods have been proposed for probabilistic reasoning using graphical models. Not all of them admit effective extension into the multiagent paradigm. Concise message passing in a compiled, treelike graphical structure has emerged from a decade's research as one class of methods that extends well into the multiagent paradigm. How to structure multiple agents' diverse knowledge on a complex environment as a set of coherent probabilistic graphical models, how to compile these models into graphical structures that support concise message passing, and how to perform concise message passing to accomplish tasks in model verification, model compilation, and distributed inference are the foci of the book. The advantages of concise message passing over alternative methods are also analyzed.

It would be impossible to present multiagent probabilistic reasoning without an exposition of its single-agent counterpart. The results from single-agent inference have been the subject of several books (Pearl [52]; Neapolitan [43]; Lauritzen [36]; Jensen [29]; Shafer [62]; Castillo, Gutierrez, and Hadi [6]; and Cowell et al. [9]). Only a small subset of these results, which were most influential to the work presented on multiagent probabilistic reasoning, is included in this book. In particular, only the theory and algorithms central to concise message-passing methods are covered in detail. These results are attributed mainly to the work of J. Pearl and his disciples as well as the Hugin researchers in Denmark. In presenting these results, instead of describing them as given solutions, the book is structured to emphasize why essential aspects of these solutions are necessary. Results from the author's own research in this regard are presented.

The book is organized into two parts. The first part includes Chapters 1 through 5 and covers probabilistic inference by concise message passing under the single-agent paradigm. Readers are prepared for comprehension of the second half of the book on multiagent probabilistic inference. The second part comprises Chapters 6 through 10 in which a formal framework is elaborated for distributed representation of probabilistic knowledge in a cooperative multiagent system and for effective, exact, and distributed inference by the agents.

Chapter 1 outlines the roles of intelligent agents and multiagent systems in decision support systems and substantiates the need for uncertain reasoning. Chapter 2 introduces Bayesian networks as a concise representation of probabilistic knowledge and demonstrates the idea of belief updating by concise message passing. Chapter 3 introduces cluster graphs as alternative models for effective belief updating by concise message passing. Through analyses of possible types of cycles in cluster graphs, this chapter formally establishes that belief updating by concise message passing requires cluster trees and, in particular, junction tree models. Chapter 4 defines graphical separation criteria in three types of graphical models and the concept of I-maps. The chapter describes stepwise how to compile a Bayesian network into a junction tree model while preserving the I-mapness as much as possible. Chapter 5 defines common operations on potentials and presents laws governing mixed operations. Algorithms for belief updating by passing potentials as messages in a junction tree are presented. Chapter 6 sets forth five basic assumptions on uncertain reasoning in a cooperative multiagent system. The logic consequences of these assumptions, which imply a particular knowledge representation formalism termed a multiply sectioned Bayesian network (MSBN), are derived. Chapter 7 presents a set of distributed algorithms used to compile an MSBN into a collection of related junction tree models, termed a linked junction forest, for effective multiagent belief updating. Chapter 8 describes a set of algorithms for performing multiagent communication and belief updating by concise message passing. The material presented in this chapter establishes that multiagent probabilistic reasoning using an MSBN is exact, distributed, and efficient (when the MSBN is sparse). Chapter 9 addresses the issues of model construction and verification and presents distributed algorithms for ensuring the integration of independently developed agents into a syntactically and semantically correct MSBN.

Chapter 10 puts the state of affairs in cooperative multiagent probabilistic reasoning in perspective and outlines several research issues in extending MSBNs into more powerful frameworks for future intelligent decision support systems.

The book is intended for researchers, educators, practitioners, and graduate students in artificial intelligence, multiagent systems, uncertain reasoning, operations research, and statistics. It can be used for self-study, as a handbook for practitioners, or as a supplemental text for graduate-level courses on multiagent systems or uncertain reasoning with graphical models. A set of exercises is included at the end of most chapters for teaching and learning. Familiarity with algorithms and mathematical exposure from a typical computer science undergraduate curriculum (discrete structure and probability) are sufficient background. Previous exposure to artificial intelligence and distributed systems is beneficial but not required.

The book treats major results formally with the underlying ideas motivated and explained intuitively, and the algorithms as well as other results are demonstrated through many examples. All algorithms are presented at sufficient levels of detail for implementation. They are written in pseudocode and can be implemented with languages of the reader's choice. The executable code of a Java-based toolkit *WebWeavr*, which implements most of the algorithms in the book, can be downloaded from

$$http://snowhite.cis.uoguelph.ca/faculty\_info/yxiang/$$

Most of the chapters (Chapters 2 through 9) contain a Guide to Chapter section as a short roadmap to the chapter. Styled differently from the rest of the chapter, this section presents no formal materials. Instead, the main issues, ideas, and results are intuitively described and often illustrated with simple examples. These sections can be used collectively as a quick tour of the more formal content of the book. They can also be used by practitioners to determine the right focus of materials for their needs.

The following convention is followed in numbering theorem-like structures: Within each chapter, all algorithms are numbered with a single sequence, and all other formal structures are numbered with another sequence, including definitions, lemmas, propositions, theorems, and corollaries.

The input, inspiration, and support of many people were critical in making this book a reality, and I am especially grateful to them: David Poole introduced me to the literature on uncertain reasoning with graphical models. Michael Beddoes made the PainULim project, during which the framework of single-agent oriented MSBNs was born, possible. Andrew Eisen and Bhanu Pant provided domain expertise in the PainULim project, and their intuition inspired the ideas behind the formal MSBN framework. Judea Pearl acted as the external examiner for my Ph.D. dissertation in which the theory of MSBNs was first documented. I owe a great deal to Bill

Havens for supporting my postdoctoral research. Nick Cercone has been a long-time colleague and has given me much support and encouragement over the years. Finn Jensen invited me for a research trip to Aalborg University during which many interesting interactions and exchanges of ideas took place. Victor Lesser was the host of my one-year sabbatical at the University of Massachusetts, and for years he has encouraged and supported the work toward a multiagent inference framework based on MSBNs. Michael Wong taught me much when I was a junior faculty member.

The work reported has benefited from my interaction with many academic colleagues, mostly in the fields of multiagent systems and uncertain reasoning: Craig Boutilier, Brahim Chaib-draa, Bruce D'Ambrosio, Keith Decker, Abhijit Deshmukh, Robert Dodier, Edmund Durfee, Piotr Gmytrasiewicz, Randy Goebel, Carl Hewitt, Michael Huhns, Stephen Joseph, Uffe Kjaerulff, Burton Lee, Alan Mackworth, Eric Neufeld, Kristian Olesen, Simon Parsons, Gregory Provan, Tuomas Sandholm, Eugene Santos, Jr., Paul Snow, Michael Wellman, Nevin Lianwen Zhang, and Shlomo Zilberstein. Students in the Intelligent Decision Support Systems Laboratory have been very helpful. Xiaoyun Chen read and commented on the drafts. I thank the users of the *WebWeavr* toolkit throughout the world for their interest and encouragement, and I hope to make an enhanced version of the toolkit available soon. My thanks also go to anonymous reviewers whose comments on the early draft proved valuable.

The Natural Sciences and Engineering Research Council (NSERC) deserves acknowledgment for sponsoring the research that has led to these results. Additional funding was provided by the Institute of Robotics and Intelligent Systems (IRIS) in the Networks of Centres of Excellence program. A significant portion of the research presented was conducted while I was a faculty member at the University of Regina. I am grateful to my many colleagues there, chaired by Brien Maguire at the time of my departure, for years of cooperation and friendship. Some of the manuscript was completed while I was visiting the University of Massachusetts at Amherst, and it was partially funded by the National Science Foundation (NSF) and the Defense Advanced Research Projects Agency (DARPA).

I would like to thank the editorial and production staffs, Lauren Cowles and Caitlin Doggart at Cambridge University Press, and Eleanor Umali and John Joswick at TechBooks for their hard work.

I am greatly indebted to my parents for their caring and patience during my extended absence. I especially would like to thank my wife, Zoe, for her love, encouragement, and support. I am also grateful to my children, Jing and Jasen, for learning to live out of a cardboard box as we moved across the country.

1

Introduction

1.1 Intelligent Agents

An *intelligent agent* is a computational or natural system that senses its environment and takes actions intelligently according to its goals. We focus on computational (versus natural) agents that act in the interests of their human principals. Such intelligent agents aid humans in making decisions. Intelligent agents can play several possible roles in the human decision process. They may play the roles of a consultant, an assistant, or a delegate. For simplicity, we will refer to intelligent agents as just *agents*.

When an agent acts as a consultant (Figure 1.1), it senses the environment but does not take actions directly. Instead, it tells the human principal what it thinks should be done. The final decision rests on the human principal. Many expert systems, such as medical expert systems (Teach and Shortliffe [75]), are used in this way. In one possible scenario, human doctors independently examine patients and arrive at their own opinions about the diseases in question. However, before the physicians finalize their diagnoses and treatments, the recommendations from expert systems are considered, possibly causing the doctors to revise their original opinions. Intelligent agents are used as consultants when the decision process can be conducted properly by humans with satisfactory results, the consequences of a bad decision are serious, and agent performance is comparable to that of humans but the agents have not been accorded high degrees of trust.

When an agent acts as an assistant (Figure 1.2), the raw data and observations are *directly* processed only by the agent. It either preprocesses the information and presents it to the human principal for further decision making or conducts the entire decision process and offers the recommendations to the human principal for approval and execution. Software systems commonly referred to as *decision support systems* (Druzdzel and Flynn [16]) are used as human assistants. For example, a corporate executive manager may use such a system to model past business data,

1

2 *Introduction*

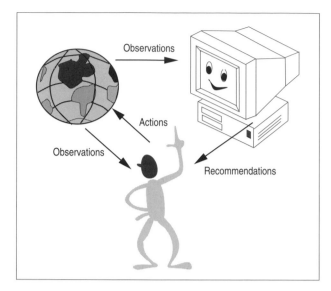

Figure 1.1: An agent as a consultant. The globe denotes the environment, and the computer denotes the agent.

analyze the consequences of alternative business actions, and arrive at a business decision. Due to the vast amount of business data and the complex interdependence among different aspects of business practice (such as material supply, production, personnel, marketing, sales, and investment), human cognitive capabilities limit

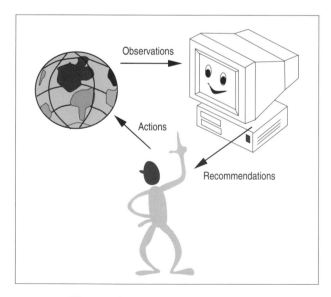

Figure 1.2: An agent as an assistant.

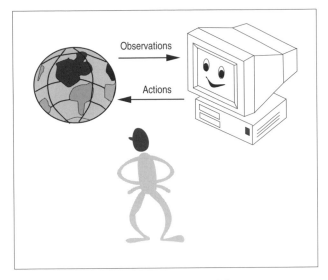

Figure 1.3: An agent as a delegate.

the manager from effectively processing all available information and arriving at a rational decision. With the aid of decision support systems, the manager only needs to evaluate preprocessed information and final or semifinal decision recommendations, which are much smaller in volume and higher in abstraction level. Intelligent agents are used as assistants when the entire decision process is beyond human cognitive capabilities given the time and decision quality constraints, for intelligent agents are capable of producing quality decision recommendations in a short time.

When an agent acts as a delegate (Figure 1.3), it not only processes the observations and generates the decision but also directly executes the chosen action without requiring approval from the human principal. Autonomous robots, Internet infobots (who search Web sites and gather relevant information for human principals), intelligent tutors, and embedded intelligent systems (such as an autodriver in a smart car) are some examples of agents being used as delegates. For example, an intelligent tutor probes the mind of a student by posing queries, determines the student's level of knowledge on the subject, detects misconceptions and weakness in skills, devises lessons and exercises to educate the student, corrects misconceptions, and strengthens the student's skills. Throughout the decision process, the tutor receives no intervention from the student's human teachers or parents (the principals of the agent). Intelligent agents are used as delegates for humans when the costs of humans' performing the tasks are too high or the tasks are too dangerous to be performed by humans and the agents are capable of high-quality performance and have won human trust.

As the information available to human decision makers continues to grow beyond human cognitive capabilities, as the cost of computing power continues to drop, and as more intelligent agents with equihuman or superhuman performance are developed, we expect to see more agents deployed as human assistants and delegates that will play increasingly important roles in daily life.

1.2 Reasoning about the Environment

Whether an agent acts as a consultant, an assistant, or a delegate, its task contains a set of essential activities. We consider agents whose activities can be described in general as follows: The agent is associated with an environment or a problem domain of interest and carries a model or a representation or some prior knowledge about the domain. Its goal is to change (or maintain) the state of the domain in some desirable way (according to the interest of its human principal). To do so, it takes actions (or recommends actions to its human principal) from time to time. To take the proper action, it makes observations on the domain, guesses the state of the domain based on the observations and its prior knowledge about the domain, and determines the most appropriate action based on its belief and goal. We refer to the activity of guessing the state of the domain from prior knowledge and observations as *reasoning* or *inference*.

An agent's activities may not fall under the preceding description. It may not separate reasoning about the state of the domain from choosing the actions but instead encodes its action directly as a function of observations. For example, an if-then rule in a rule-based expert system may have its *if* part specifying some observations and its *then* part specifying a desirable action when the observations are obtained. An if-then rule is made out of symbols (observations and actions are specified in terms of symbols). Hence, an agent constructed from a rule-based system uses a symbolic (Poole, Mackworth, and Goebel [55]) knowledge representation of the domain. An agent may not even use a symbolic representation. For example, an agent's behavior may be based on an *artificial neural network* (Hassoun [23]) in which the observations are mapped to proper actions through network links and hidden units that do not have well-defined semantics.

In this book, we consider agents using symbolic knowledge representations and reasoning explicitly about the state of the domain. Reasoning about the state of the domain can be a challenging task by itself, as we shall see. Separating the reasoning task from the selection of actions allows decomposing the decision process into two stages and working on each through divide-and-conquer techniques. To develop agents with quality performance, one approach is to ensure that agents' behavior can be analyzed formally and rigorously. Separating inference about the state from choice of action also facilitates the analysis and explanation of agents' behavior by analyzing and understanding each stage of the decision process individually.

1.3 Why Uncertain Reasoning?

The main focus of this book is on guessing (reasoning about) the state of the domain. We have used the word *guessing* for several reasons.

1. Some aspects of the domain are often unobservable and therefore must be estimated indirectly through observables. Consider a neurologist who needs to know which part of an epileptic brain is abnormal. Direct examination of the brain is not an option except as part of the surgical procedure after the diagnosis has been completed and surgery has been considered necessary. Instead, to estimate the state of the brain, the seizure behavior of the patient is observed, and the electroencephalogram (EEG) is recorded.

    Complex equipment, such as automobiles, airplanes, automatic production lines, chemical plants, computer systems, and computer networks, is used widely in modern society. Each piece of equipment is made of many components, each of which is further composed of many devices and parts. Whether a device or component is functionally normal or faulty or on the verge of breaking down is usually not observable. For example, whether the bearing of a helicopter propeller is about to break down is not directly observable without disassembling the propeller, but such knowledge is crucial to prevent accidents. To estimate the state of the bearing, sensors are often used to collect the vibration patterns of the bearing.

    In making financial and economic decisions as corporate executives and government officials do, it is often advantageous to take the upcoming economic climate into account. Is a recession or an economic boom on the horizon? Although this process is more concerned with predicting the future than evaluating current conditions, the trend can be viewed as one characteristic of the present economic state that is not directly observable.

    In playing card games such as poker, knowing the opponent's hand allows a player to determine his or her best strategy with certainty. Because the opponent's hand is not observable, the best one can do is to guess it based on the cards revealed so far and the past behavior of the opponent.

2. The relations among domain events are often uncertain. In particular, the relationship between the observables and nonobservables is often uncertain. Therefore, the nonobservables cannot be inferred from the observables with certainty. The uncertain relation prevents a reasoner from arriving at a *unique* domain state from observation, which accounts for the guesswork. For instance, certain seizure behavior can be caused by lesions at different locations of the brain, which makes it difficult to determine exactly where surgery should take place.

    Most equipment is intended to work deterministically. Given the input of a piece of equipment or, in general, the input history, a unique, desirable response or response sequence is expected. For example, when the brake pedal is pressed, a running car should slow down. After an enlarging scale has been entered into a photocopier, pressing the start button should cause the lens to reposition accordingly before photocopying is performed. However, equipment may fail due to failure of components and devices. Because failure is normally not designed, the faulty behavior of a piece of equipment is generally uncertain. The break down of different devices or parts of a piece of equipment may cause the same failure mode. On the other hand, the same faulty device may generate different failures owing to the mode of the equipment, the raw material or input being processed at the time, and other factors. A somewhat worn-out propeller bearing may last quite a while if the helicopter carries light loads and flies in good weather conditions, but the bearing may break down much sooner and cause an accident if the aircraft flies in severe weather with heavy loads.

Earthquakes are often preceded by indicators such as sudden changes in water well levels, abnormal behavior of domestic animals, and instrumental indications. On February 4, 1975, a major earthquake (Holland [25]) struck a heavily populated area 400 miles northeast of Beijing, China. Ninety percent of the buildings in some areas were destroyed as was the entire town of Haicheng. On the basis of the indicators, a prediction was made, and nearly one million people who lived in the area were evacuated just hours before the earthquake. As a result, no one was killed when the earthquake struck. However, reliable predictions of major earthquakes are not yet a reality, and successful predictions such as this one have been rare. Consequently, disasters due to unpredicted major earthquakes do occur from time to time.

3. The observations themselves may be imprecise, ambiguous, vague, noisy, and unreliable. They introduce additional uncertainty to inferring the state of the domain. The EEG recorded from a patient cannot reflect the electromagnetic activities that are far from the scalp. On the other hand, the signals recorded from each electrode are the summation of the activities of many neurons (some normal and some lesioned) in the nearby brain plus artifacts (e.g., due to electrode movement or muscle activity).

To monitor complex equipment, sensors are commonly used to extract information about the temperature, pressure, displacement, altitude, speed, vibration, and other physical quantities from the components or devices. The states of the components and the equipment are then inferred from the sensor outputs. However, sensors may not respond to the target quantities evenly under different conditions and hence can introduce errors. Sensors can pick up signals from nearby sources in addition to the target sources. Sensors can fail and consequently produce outputs correlated or uncorrelated to the source quantities. Sensor outputs can be contaminated by noise while being transmitted from the sensors to the processing units. For example, if the helicopter bearing vibration is monitored by a sensor and the sensor output is transmitted to a unit located in the cockpit, what is received by the unit may contain other nearby electromagnetic signals.

When a witness testifies in court and states that he or she saw a suspect on site from a distance on the night of a murder, the reliability of this statement is to be judged by taking into account the illumination of the scene, the distance between the witness and the suspect at the time, the vision of the witness, and the relation between the witness and the suspect. Similarly, when asking for direction in an unfamiliar area, if you are told that your destination is "not too far," a wide range of distance is still possible.

4. Even though many relevant events are observable, very often we do not have the resources to observe them all. Therefore, the state of the domain must be guessed based on incomplete information. In medicine, even though many laboratory tests may help improve the accuracy of a patient's diagnosis, not all such tests will be performed on a

patient due to the cost involved and the potential side effects. Hence, the diagnosis must be made based on the routine physical examination and limited laboratory tests.

On battlefields, gathering more intelligence reports can result in more accurate knowledge about the enemy's location, movement, and intention and hence can lead to a more informed strategy. Because of the time and risk involved in gathering these reports, the enemy's state must very often be guessed using only limited reports.

5. Even though event relations are certain in some domains, very often it is impractical to analyze all of them explicitly. Consequently, the state of the domain is estimated from computationally more efficient but less reliable means. In board games, the configuration is certain, and the consequences of all legal moves of the player and the opponent are also certain. However, for many board games, due to the huge number of combinations of legal moves it is not feasible to analyze each of them (to an endgame) before making the current move. As a result, less reliable but more efficient heuristic functions are used to evaluate each board configuration and potential move.

A mechanical workshop manufacturing parts on contract needs to schedule which machine processes which part at which time slot. As the number of machines, the number of different parts to be manufactured, and the operations to be performed on each part increase, finding the optimal schedule for manufacturing all parts in the shortest time becomes impracticable.

In the light of these factors and others, the reasoner's task is not one of deterministic deduction but rather uncertain reasoning. That is, the reasoner must infer the state of the domain based on incomplete and uncertain knowledge about the domain and incomplete and uncertain observations. Many competing theories exist on how to perform such uncertain reasoning. This book focuses on methodologies founded on Bayesian probability theory. In other words, it focuses on *probabilistic reasoning*. A body of literature exists that compares the relative merits of alternatives.

1.4 Multiagent Systems

We have considered an agent that makes observations on a domain, performs probabilistic inference based on its knowledge about the relations among domain events, and estimates the state of the domain. However, a single agent is limited by its knowledge, its perspective, and its computational resources. As the domain becomes larger and more complex, open, and distributed, a set of cooperating agents is needed to address the reasoning task effectively.

Imagine a smart house (Boman, Davidsson, and Younes [4]), which is likely to be available in the near future. Compared with existing houses, the appliances and other components in a smart house are operated more energy efficiently and provide better comfort to the occupants. Curtains are closed automatically in the evening when it becomes dark, and lights are turned on. In the winter, heating is reduced when people are out working and is restored to the previous setting before the occupants return. Air conditioning is handled similarly on hot summer days. The temperature in each room is adjusted individually according to the preference of the occupant(s) and the activity conducted. Perhaps the room for fitness is kept cooler than the family room. The refrigerator sends an order to the local grocery supplier through the Internet whenever the supply of milk and eggs is low. The oven starts cooking in the late afternoon before family members are home so that the dinner will be ready at the right time. The sprinkler system turns on after several dry summer days but will save the water when it rains from time to time.

The proper operation of these components depends on believing that certain events have happened or are about to happen such as "people are out working," "people will return shortly," "the grocery supply is low," and "it's been dry for quite a while." Knowledge and reasoning generate such beliefs. Consider estimating the occurrence of "people are out working." A simple timing based on a rigid schedule is not sufficient because one family member may be sick at home on a certain day and not follow the regular schedule. Motion detectors are not foolproof either. A dog may be wandering around the house when no one else is home. The dog may cause the motion detector to believe that a family member is home; hence, heating will not be reduced as expected. A patient may have caught the flu and be in bed without much motion during the day. Because no motion is detected and no one is believed at home, heating is reduced. This may worsen the patient's condition. Prior knowledge about each family member's normal work schedule, the expected activities during sickness, the existence of a pet and its behavior, as well as outputs from different sensors can all contribute, through reasoning, to the belief "people are out working." The available prior knowledge is generally uncertain. For example, how much remaining grocery supply constitutes a "low" supply depends on the eating habits of the family, the day of the week (people may eat differently on weekdays than weekends), and other factors.

The proper operation of these components also depends on the knowledge of, and belief in, the functionality and expected behavior of appliances. When to turn off the sprinkler depends on determining whether the lawn has been watered sufficiently. This in turn depends on the knowledge of the sprinkler system's capacity and the size of the lawn and the belief about the lawn's degree of dryness. How early to start cooking before the family is back from work depends on the belief about what is to be cooked and knowledge of how long it takes the oven to do the cooking. Clearly, diverse knowledge about household components is needed. With new appliances installed or upgraded, an open-ended set of knowledge needs to be managed and maintained.

Often, activities in different sections of the house need to be coordinated. After the dishwasher in the kitchen is loaded and ready to wash, it may be better to delay dish washing if it is believed that a family member will start taking a shower upstairs in the washroom. If both activities are going forward, it is likely that the hot water will run out before the person finishes the shower. Such coordination can be achieved if relevant events occurring in different areas of the house are assessed properly.

Building and maintaining a single intelligent agent to manage such complex, distributed, and open-ended activities would be very costly if not impossible. An alternative to the single-agent paradigm is the paradigm of *multiagent systems*. A multiagent system consists of a set of agents acting in a problem domain. Each agent carries only a partial knowledge representation about the domain and can observe the domain from a partial perspective. Although an agent in a multiagent system can reason and act autonomously as in the single-agent paradigm, to overcome its limit in domain knowledge, perspective, and computational resource, it can benefit from other agents' knowledge, perspectives, and computational resources through communication and coordination. This multiagent paradigm is promising for overcoming the limitation of the single-agent paradigm, as discussed in the following paragraphs.

1. In large and complex domains, diverse knowledge is required. In the smart house domain, the knowledge of different household appliances and components and of human activities and behaviors is needed to operate intelligently. A powerful tool to handle such complexity and diversity is *modularity*. Under the multiagent paradigm, for each appliance or component we can construct an agent capable of operating the unit. Because such an agent requires only limited knowledge, this approach simplifies development. The interdependence between units is handled by coordination among agents.

   As a different example, consider equipment monitoring and diagnosis (M&D). The total complexity of a piece of complex modern equipment (e.g., an airplane or a chemical plant) is usually beyond the comprehension of a single person or even a single team. One reflection is that it is increasingly common for the manufacturer of a particular piece of equipment to purchase half or even more of the components from other vendors (Parunak [47]). Let R be a manufacturer who needs a component c for its product and S be the supplier of c. Then R must have the knowledge of how c should function in terms of its input–output relation so that R can integrate c with other components – purchased or manufactured. However, more detailed knowledge about the internals of c, which is necessary to monitor and diagnose c, may not be available from S. Even if it is available, R may not want to bother with it in order to manage its core business more efficiently. In such a case, to build an M&D system, R may instead use an M&D agent for c developed by S and let it cooperate with agents responsible for other components.

2. In large and complex domains, sensors are often distributed. In a smart house, sensors can collect data on temperature, humidity, object movement, lighting, water usage, and other events in each room and near the house. Components in a complex system are often physically distributed (e.g., heaters and compressors in a chemical plant). Sensors to collect

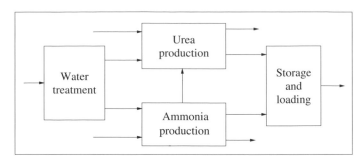

Figure 1.4: Major plants in a fertilizer factory.

observations are therefore also distributed. Figure 1.4 shows a top-level decomposition of a fertilizer factory consisting of four plants. Each plant can be further decomposed into several major components. Figure 1.5 shows the decomposition of the water treatment plant. A large number of sensors are normally placed throughout the factory to collect observations about components.

Traditionally, the observations from distributed sensors are transmitted to a central place. There, they are processed by a single agent, the necessary actions are decided, and the control signals are transmitted to the locations where the actions take place. Transmission of the observations and action control signals, however, is limited by the available bandwidth, the time delay, and potential interruption due to failure of the communication channels. The multiagent paradigm suggests deployment of multiple agents, each near a component or a small group of nearby components, so that the sensor outputs can be processed on site and actions be taken more promptly (e.g., opening a local valve to release pressure believed to be beyond the safety level).

3. Many complex problem domains are open-ended. Each smart house will have a different set of appliances and components in accord with the family's need and budget. As time goes by, new appliances may be installed and existing appliances may be replaced or upgraded. Similar situations happen with equipment. New functional components may

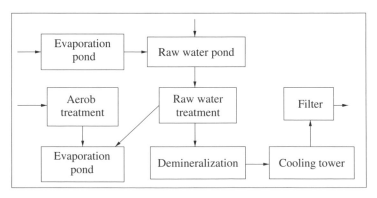

Figure 1.5: Major equipment in a water treatment plant.

be added as the need arises or as the vendors release them. By capturing the knowledge and decision making relative to each component into a separate agent, the addition of new components is handled in a multiagent system by dynamically adding agents and letting existing agents adapt to the new agent community.

An important feature of multiagent systems is that agents in such systems are *autonomous*. The autonomy is reflected in the interrelation of agents and their human principals as well as between agents. Due to the large number of agents in a multiagent system, it is impractical for a human principal to guide each agent closely during its activities. Although the entire system can still play the three possible roles (consultant, assistant, and delegate), individual agents within the system must be able to reason and act on their own most of the time with minimal human intervention. In other words, a centralized intervention is mostly unavailable. Furthermore, because each agent is intended to process a locally accessible information source and to solve a partial problem based on its local computational resource, it follows that communication between agents will be concise and infrequent. In other words, constant and raw-data-based communication among agents is mostly unavailable. As will be seen in this book, these implications of autonomy exert significant constraints on the design choices of multiagent systems.

Research and practice using multiagent systems are closely related to *object-oriented programming* in which an object encapsulates its state, can only be accessed and modified by its own methods, and forms a unit in software design; are closely related to *distributed systems* research through which hardware, software, and network structures for fast and reliable communication and efficient distributed computation are developed; and are closely related to *human–computer interface* research in which task delegation is used as an alternative to direct manipulation. Sycara [71] discusses one set of criteria with which the relation and difference between these (and related) research and multiagent systems can be identified. In this book, we take a relatively loose notion of agents reminiscent of Poole et al. [55] and hence a loose notion of multiagent systems. In the next section, we introduce the task of probabilistic reasoning by multiple agents. In the later chapters of the book, we precisely define the task of multiagent probabilistic reasoning and study how such a task can be performed.

1.5 Cooperative Multiagent Probabilistic Reasoning

In our discussion of single-agent decision making, we decomposed the agent's decision process into reasoning about the domain state and selection of actions. We will apply a similar decomposition to a multiagent system and study the subtask of how multiple agents can collectively reason about the state of the domain based on their local knowledge, local observation, and limited communication. This subtask is referred to by some authors as *distributed interpretation* (Lesser and Erman [38]). As can be imagined, the needs and the opportunities of communication lead to additional issues for multiagent reasoning.

What is the objective of communication in a cooperative multiagent system? Should agents exchange their observations or their beliefs? If each agent has only a partial perspective of the domain, what should be the relationship between their beliefs? For example, should agents be allowed to hold inconsistent beliefs? Is there such a thing as the collective belief of multiple agents? If so, what form does it take?

If communication is to restore belief consistency among agents, how should the communication be structured and organized? Whom should an agent communicate with? Should an agent be allowed to communicate with any other agent? What is the consequence of free communication with respect to the objective of agent communication?

What information should an agent exchange with other agents? Too much information is unnecessary and inefficient. Too little information does not benefit from the full potential of communication. How do we determine the right amount of information to be exchanged?

In data communication, data can be compressed and then transmitted so that the same amount of data can be communicated with less bandwidth and channel time. Such channel coding can be applied at different levels of abstraction (e.g., at bit level or at word level). How can the information to be exchanged between agents be efficiently encoded at the knowledge level?

In building complex systems, there is always the trade-off between system performance and system complexity required to deliver that performance. For agents to communicate effectively and believe rationally, a certain structure and organization may be necessary. Once such structure and organization are identified, how can agents constructed by independent vendors be integrated into a multiagent system that respects the structure and organization? How can the structure and organization be verified without violating agent privacy (and ultimately protecting the proprietary information and technical know-how of the agent vendor)? We will study these issues in this book.

Each agent in a multiagent system may serve the interest of a different principal. Because the human principals may have conflicting interests (such as a business whose main concern is profit and a customer whose main concern is the quality of goods and service purchased), agents that serve multiple principals of conflicting interests are called *self-interested* agents. If all agents in a multiagent system serve the interest of a single principal (which could be a human organization), they are called *cooperative* agents.

The communications behaviors of cooperative versus self-interested agents are quite different. Although cooperative agents can be assumed to be truthful to each other because they are working for a common principal, a self-interested agent may deliberately provide false information to other agents who serve different principals. This difference in agent communications behavior implies that assumptions, principles, and techniques applicable to cooperative multiagent reasoning may not be applicable in general to reasoning of self-interested agents. The focus of this book is on cooperative agents and, in particular, on probabilistic reasoning of cooperative agents. Readers who are interested in inference and decision making in systems consisting primarily of self-interested agents are directed to references such as Rosenschein and Zlotkin [59] and Sandholm (Chapter 5 in Weiss [77]).

1.6 Application Domains

Not all complex domains are suitable for *cooperative* multiagent systems. Depending on the degree and nature of the uncertainty in the domain and its impact on the quality of decision, uncertain reasoning may not be a significant component in a cooperative multiagent system. For example, Tacair–Soar (Jones et al. [31]), a large-scale combat flight simulation system, does not explicitly perform uncertainty reasoning. As explained by a Tacair–Soar team member, their system was developed in this way because the worst-case scenario is usually the basis for combat pilot decision making.

Nevertheless, uncertain problem domains suitable for cooperative multiagent systems are abundant. Many of them involve a nontrivial subtask of estimating the current state of the domain to facilitate action formulation. We have mentioned monitoring and diagnosis of complex equipment and processes as well as smart houses. These domains are examples of general sensor networks for surveillance, monitoring, hazard prediction and warning for buildings, warehouses, restricted areas, computer networks, and industrial processes. In business domains, monitoring and interpretation of corporate operating status is an important subtask in management decisions. In distributed design, whether design choices made at different components by diverse designers lead to a system of desirable performance that takes into account many uncertain factors in materials, manufacturing, operation, and maintenance can also be treated as a problem of cooperative uncertain reasoning.

Cooperative multiagent probabilistic reasoning in complex domains is a nontrivial task, as we will see. The general technical issues involved are better comprehended when illustrated with examples. However, examples for sophisticated technical domains demand significant background domain knowledge from the readers, which hinders reader comprehension. To avoid such a burden, we choose digital electronic systems as the source of examples when a large problem domain is needed. Digital electronic systems are suitable for this purpose for the reasons discussed next.

Although a domain of about 20 variables may be sufficient to illustrate many issues involved in modeling a single agent, it will be too small to illustrate issues involved in modeling a multiagent system. On the other hand, comprehension of technical details of an example from a specialized domain with a reasonable size (e.g., the fertilizer factory in Figure 1.4) demands an unreasonable amount of background knowledge from readers, which distracts them from the general issues in question. The compromise made is to construct large examples from a domain of knowledge common to most readers. A basic understanding of digital electronics can safely be assumed for all professionals in information technology and for many in science and engineering. Perhaps this is one of the major reasons why digital electronics has been the source of problems for many researchers in diagnosis (e.g., Davis [11], Genesereth [20], de Kleer and Williams [13], Pearl [52], Poole [54], and Srinivas [69]).

As with any other equipment, a digital system is intended to work deterministically, but the failure behavior is uncertain. Hence, the use of a digital domain does not diminish the number of general issues related to uncertain reasoning. Furthermore, the complexity in modeling and inference using probabilistic graphical models grows as the degree of network nodes increases and the number of loops in the network increases. In a digital system, the former corresponds to the number of inputs and outputs for a particular gate or device, and the latter is reflected in the circuit topology.

A digital system may be *combinatorial* or *sequential*. In a combinatorial circuit, output values depend on only the input values, whereas in a sequential circuit output values depend also on the internal state of the circuit which is determined by the history of inputs. Therefore, a combinatorial circuit system provides a *static* domain, whereas a sequential circuit system provides a *dynamic* domain. Hence, issues on diagnosis in both static and dynamic domains can be illustrated properly using digital electronic systems.

1.7 Bibliographical Notes

Motivations for uncertain reasoning in intelligent systems can be found in several recent artificial intelligence textbooks including Russell and Norvig [60]; Dean, Allen, and Aloimonos [14]; and Poole et al. [55]. A collection edited by Shafer and Pearl [63] presents a number of alternative approaches to uncertain reasoning. The February 1988 issue of *Computational Intelligence* journal contains a lively debate over alternative approaches for commonsense reasoning with a position paper by Cheeseman [7] and comments by 20 authors. Limitations of human reasoning under uncertainty are studied in Kahneman, Slovic, and Tversky [32].

The notion of agents has been adopted by all recent artificial intelligence textbooks (Russell and Norvig [60], Dean et al. [14], Poole et al. [55], and Nilsson [44]). Introductions to multiagent systems can be found in Wooldridge and Jennings [79] and Sycara [71]. Earlier multiagent system research is covered in Bond and Gasser [5], and more recent advances are contained in a comprehensive collection edited by Weiss [77]. Reasoning and decision making for self-interested agents are studied in Rosenschein and Zlotkin [59] and Sandholm (Chapter 5 in Weiss [77]).

2

Bayesian Networks

To act in a complex problem domain, a decision maker needs to know the current state of the domain in order to choose the most appropriate action. In a domain about which the decision maker has only uncertain knowledge and partial observations, it is often impossible to estimate the state of the domain with certainty. We introduce Bayesian networks as a concise graphical representation of a decision maker's probabilistic knowledge of an uncertain domain. We raise the issue of how to use such knowledge to estimate the current state of the domain effectively. To accomplish this task, the idea of message passing in graphical models is illustrated with several alternative methods. Subsequent chapters will present representational and computational techniques to address the limitation of these methods.

The basics of Bayesian probability theory are reviewed in Section 2.2. This is followed in Section 2.3 by a demonstration of the intractability of traditional belief updating using joint probability distributions. The necessary background in graph theory is then provided in Section 2.4. Section 2.5 introduces Bayesian networks as a concise graphical model for probabilistic knowledge. In Section 2.6, the fundamental idea of local computation and message passing in modern probabilistic inference using graphical models is illustrated using so-called $\lambda - \pi$ message passing in tree-structured models. The limitation of $\lambda - \pi$ message passing is discussed followed by the presentation of an alternative exact inference method, loop cutset conditioning, in Section 2.7 and an alternative approximate inference method, forward stochastic sampling, in Section 2.8. The behaviors of the alternative methods with respect to message passing are characterized, which motivates the inference methods to be presented in subsequent chapters.

2.1 Guide to Chapter 2

Consider an intelligent agent whose job in part is to monitor a digital circuit (Figure 2.1) in an appliance. The agent needs to know whether the circuit functions

16

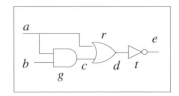

Figure 2.1: A simple digital circuit.

normally or abnormally and which devices are faulty when it functions abnormally. The state of each device is not directly observable and can only be inferred from observation of the inputs and outputs of gates. Because observations are costly (the effort to collect them and the time and bandwidth to transmit them to the agent), it is desirable not to have to observe all inputs and outputs for all gates. How should the agent go about doing its job?

We can let the agent carry a representation of its knowledge or belief about the expected behavior (both normal and abnormal) of the circuit. It can then use this knowledge plus any observation to infer the current state of the circuit. Because the knowledge is uncertain (e.g., about the abnormal behavior of the circuit), it can be coded as a probability distribution. Section 2.2 reviews the basic concepts of probability, describes the conditions that govern what is or is not a valid probability distribution, and presents several useful rules for manipulating probability. As any time, the probability distribution reflects the current belief of the agent about the state of the circuit. A *prior* probability distribution reflects the agent's belief before any observation is made, and a *posterior* probability distribution reflects the agent's belief after the observation. *Belief updating* refers to the computation carried out by the agent to update its prior belief to its posterior belief.

For the digital circuit, the agent can encode its prior belief as a probability distribution $P(a, b, c, d, e, g, r, t)$, where each symbol represents the input of a gate (e.g., b), or the output of a gate (e.g., e), or the state of a gate (e.g., g). For example,

$$P(a=0, b=0, c=0, d=0, e=1, g=\text{normal}, r=\text{normal}, t=\text{normal})=0.2$$

represents the agent's belief that 20% of the time the inputs a and b are 0, all gates are normal, and the outputs c and d are 0 and the output e is 1. On the other hand,

$$P(a=0, b=0, c=0, d=0, e=0, g=\text{normal}, r=\text{normal}, t=\text{abnormal})$$
$$= 0.009$$

represents the agent's belief that 0.9% of the time all other values are the same as those given above but the gate t is abnormal and the output e is 0. The agent's belief about any gate such as g can also be derived from the preceding probability distribution in the form of a probability distribution $P(g)$. The procedure for performing such derivations is shown in Section 2.2.

Section 2.3 describes how belief updating can be performed by updating the prior $P(g)$ to the posterior $P(g|a = 0, b = 0, e = 0)$ when a, b, and e are observed. It also demonstrates that, as the number of gates in the circuit increases, representation and belief updating based on such representations become intractable. Coping with these intractabilities is the primary concern of this book.

This leads to the idea of encoding the probabilistic knowledge concisely using graphical models and effectively updating belief with these models. A *graph* is made of *nodes* and *links* that connect them. We can use nodes to represent events and links between a pair of nodes to represent the relation between them. To do so, we first need to be able to describe the topology of a graph. Section 2.4 introduces the basic concepts and terminologies for graphs.

Graphs can be of different types such as *directed* (where links in the graph have directions), *undirected* (where links have no directions), and *hybrid*. For undirected graphs, terminologies are introduced to describe the neighborhood of a node, a *path* from one node to another, a *cycle* in a graph, a graph within another graph, and other aspects of connectivity of a graph. For directed graphs, additional terminologies are introduced to describe a directed path or cycle and to name nodes at different locations on a directed path.

Section 2.5 relates the probabilistic knowledge of an agent to a graphical model called a *Bayesian network* (BN). Such a network is a directed graphical model. Each node in the graph signifies an event. Each directed link represents a causal dependence between two events. Figure 2.2 shows the graph of a BN representing an agent's knowledge of the digital circuit. How to construct such a graph and how to encode the agent's probabilistic knowledge concisely using the graph will be discussed in Section 2.5. The way in which BNs help cope with the representation intractability mentioned earlier is also shown.

Section 2.6 addresses the belief updating intractability stated earlier. It introduces the idea of message passing along arcs of BNs as a means of effective belief updating. A method called $\lambda - \pi$ message passing is used to illustrate how

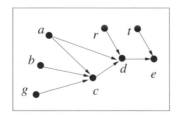

Figure 2.2: The graphical model for the digital circuit.

belief updating can be performed by passing concise messages in treelike BNs. Section 2.7 presents how $\lambda - \pi$ message passing can be applied to nontree BNs by first converting a nontree BN into multiple treelike BNs. The method is called *loop cutset conditioning*. Section 2.8 presents an approximate method, *forward stochastic sampling*, for belief updating by passing massive (versus concise) messages in a nontree BN directly.

2.2 Basics on Bayesian Probability Theory

An agent needs to know the actual state of a problem domain or parts of the domain. However, it can only directly observe some events of the domain and infer the unobservables from the observables. Suppose a technician needs to find out which components are responsible for an abnormal chemical process. Detailed examination of every component is not feasible. The technician must use the observed events to help focus on a small number of components as the candidates for the culprit.

Formally, a domain is described in terms of a set of variables $V = \{v_1, \ldots, v_n\}$. We consider only discrete variables in this book, but in general these variables may be discrete or continuous. To make the discussion concrete and manageable, consider the task of monitoring and trouble-shooting a simple digital circuit in Figure 2.3. To determine the state of the circuit, the domain is modeled by $V = \{g_1, g_2, g_3, g_4, a, b, c, d, e, f, h\}$, which consists of variables describing the state of each gate (e.g., g_1) as well as its digital input and output (e.g., b and e).

Each variable v_i takes its value from a finite *space* D_{v_i} of possible values, namely, $v_i \in D_{v_i}$. For example, the space for g_1 is $D_{g_1} = \{\text{norm, ab}\}$, where "norm" stands for *normal* and "ab" stands for *abnormal*. The space for the signal a is $D_a = \{0, 1\}$.

For a subset $X \subset V$, its space D_X is the Cartesian product of the spaces for the variables in X, namely, $D_X = \prod_{x \in X} D_x$. For instance, the space for $X = \{g_1, e\}$ is

$$D_X = \{(g_1 = \text{norm}, e = 0), (g_1 = \text{norm}, e = 1), (g_1 = ab, e = 0),$$
$$(g_1 = ab, e = 1)\}.$$

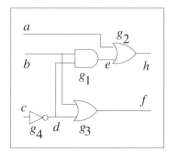

Figure 2.3: A simple digital circuit.

Each element of D_X (denoted by x) is called a *configuration* of X. A configuration x represents an *event*, namely, that each variable in X takes the value specified in x. If the event is known to be true, it is denoted as $X = x$. For example, $X = (g_1 = \text{norm}, e = 0)$ signifies that the gate g_1 is normal and its output is 0. For the problem domain V, each configuration v describes a possible state of the domain. For example, the configuration

$$(g_1 = \text{norm}, g_2 = \text{norm}, g_3 = \text{norm}, g_4 = \text{norm}, a = 0,$$
$$b = 0, c = 0, d = 1, e = 0, f = 1, h = 0)$$

describes a possible state of the domain, which, for simplicity, we sometime write as

$$(\text{norm}, \text{norm}, \text{norm}, \text{norm}, 0, 0, 0, 1, 0, 1, 0).$$

The configuration

$$(\text{norm}, \text{norm}, \text{norm}, \text{norm}, 0, 0, 0, 1, 0, 1, 1)$$

(same as above except $h = 1$) describes a state that is *impossible*.

Let X and Y be two subsets of V and x and y be their configurations. Configurations x and y are *compatible* if for every $v \in X \cap Y$, its values in x and in y are identical. The two configurations of V above are incompatible. The configurations $(g_1 = \text{norm}, b = 0)$ and $(g_3 = \text{ab}, b = 0)$ are compatible.

To infer the state of the domain from observation, an agent must rely on the constraints or dependence between observables and unobservables. For example, it can infer the state of OR gate g_2 from observations on its inputs a and e and its output h. Very often, the dependence is not deterministic but rather is uncertain. For example, when g_2 is at the state ab, its output h may or may not be 0 when both inputs are 0.

The Bayesian probability theory tells us that an agent's knowledge about the uncertain dependence between variables can be represented in terms of a probability function P defined over these variables. Hence $P(X = x | Y = y)$ represents the (degree of) *belief* of a reasoner on the truth of x when he or she already knows that y is true. It is called the *conditional probability* of x given y. The expression $P(X|Y)$ represents the beliefs for each combination of x and y and is referred to as a *conditional probability distribution* of X given Y. For simplicity, we write $P(x|y)$ instead of $P(X = x | Y = y)$. Given $X = \{h\}$ and $Y = \{a, g_2\}$, we write $P(X|Y)$ as $P(h|a, g_2)$ instead of $P(\{h\}|\{a, g_2\})$, and we write $P(h|a = 0, g_2 = \text{norm})$ instead of $P(h|Y = (a = 0, g_2 = \text{norm}))$.

An agent's knowledge of the normal behavior of g_2 when both inputs are 0 can be represented as $P(h = 0 | a = 0, e = 0, g_2 = \text{norm}) = 1$. If it is known that when g_2 is abnormal the gate will output correctly 70% of time when both inputs

Table 2.1: *Conditional probability distribution* $P(h|a, e, g_2)$

h	a	e	g_2	$P(h|a, e, g_2)$	h	a	e	g_2	$P(h|a, e, g_2)$
0	0	0	norm	1.0	0	0	0	ab	0.7
0	0	1	norm	0	0	0	1	ab	0.6
0	1	0	norm	0	0	1	0	ab	0.6
0	1	1	norm	0	0	1	1	ab	0.2
1	0	0	norm	0	1	0	0	ab	0.3
1	0	1	norm	1.0	1	0	1	ab	0.4
1	1	0	norm	1.0	1	1	0	ab	0.4
1	1	1	norm	1.0	1	1	1	ab	0.8

are 0, this can be represented by $P(h = 0|a = 0, e = 0, g_2 = \text{ab}) = 0.7$. A possible distribution for $P(h|a, e, g_2)$ is shown in Table 2.1. Because each variable is binary, the distribution has $2^4 = 16$ entries.

Sometimes, for simplicity, the background knowledge of the reasoner is not explicitly represented within V. We then have the *unconditional* probability $P(x)$ and unconditional distribution $P(X)$. The unconditional distribution $P(V)$ over the entire domain is referred to as the *joint probability distribution* (JPD). When it is emphasized that X is a subset of V or is a subset of some set Y such that $X \subset Y \subset V$, $P(X)$ is referred to as a *marginal* distribution of $P(V)$ (or $P(Y)$).

Although P can be specified subjectively, it must satisfy a set of axioms (Eqs. (2.1) through (2.4)). Let X, Y, Z be any nonempty subsets of V. Equation (2.1) says that the probability of any event is a real number between 0 and 1:

$$0 \leq P(x|y) \leq 1, \qquad (2.1)$$

where $x \in D_X$ and $y \in D_Y$. Equation (2.2) stipulates that certainty is represented by 1:

$$P(x|x) = 1, \qquad (2.2)$$

where $x \in D_X$. The following *sum* rule says that, if x and y are *mutually exclusive*, then the probability that either one will occur is the sum of the individual probabilities:

$$P(x \text{ or } y|z) = P(x|z) + P(y|z), \qquad (2.3)$$

where $x \in D_X$, $y \in D_Y$, and $z \in D_Z$. Equations (2.2) and (2.3) imply that a probability distribution sums to 1:

$$\sum_{x \in D_X} P(x|Y) = 1.$$

The following *product* rule says that the probability of the joint event that both x and y occur is the probability of one event multiplied by the probability of the other event conditioned on the first:

$$P(x, y|z) = P(x|y, z)P(y|z), \tag{2.4}$$

where $x \in D_X$, $y \in D_Y$, and $z \in D_Z$. In the equation, the comma "," represents logic conjunction "and." For simplicity, we write

$$P(X, Y|Z) = P(X|Y, Z)P(Y|Z).$$

We will adopt the simple notation whenever there is no confusion.

Many useful properties of probability distributions can be proven from the preceding axioms. One such property is *Bayes*'s rule:

$$P(X|Y, Z) = \frac{P(Y|X, Z)P(X|Z)}{P(Y|Z)}. \tag{2.5}$$

Another property is the *negation* rule:

$$P(X \neq x|y) = 1 - P(X = x|y), \tag{2.6}$$

where $x \in D_X$ and $y \in D_Y$. We often need to derive the distribution over X from the distribution over $Y \supset X$. This is what the *marginalization* rule allows us to do:

$$P(x|z) = \sum_{w \in D_W} P(x, w|z), \tag{2.7}$$

where $X \subset Y$, $W = Y \backslash X$, $x \in D_X$, and $z \in D_Z$. The operator $\backslash$ denotes *set difference*. For simplicity, we write

$$P(X|Z) = \sum_W P(X, W|Z) = \sum_W P(Y|Z),$$

and the subset W of variables is said to be *marginalized out*.

The purpose of representing knowledge over a problem domain V using a probability distribution is to be able to reason about the state of the domain given some observations. The distribution $P(X)$ on a subset $X \subset V$ of variables *before* making any observation is referred to as the *prior* distribution. After y is observed for $Y \subset V$, the distribution $P(X|Y = y)$ (or simply $P(X|y)$) is referred to as the *posterior* distribution. A fundamental task in probabilistic reasoning called *belief updating* is to update a prior distribution into a posterior distribution when observations are available.

2.3 Belief Updating Using JPD

Given a problem domain V, in principle an agent can reason about the domain using the JPD $P(V)$. Let's consider each step involved:

First, the agent needs to acquire the parameters of $P(V)$. Denote the cardinality of V by $n = |V|$ and the cardinality of the largest variable space by $d = \max_v |D_v|$. Then the agent needs to acquire $O(d^n)$ probability values to specify $P(V)$ fully. The expression $O(d^n)$ means that d^n values (or a constant time of those) are to be acquired in the *worst* case. As the cardinality of V increases, this will be an intractable task. We refer to this difficulty as *acquisition intractability*.

Next, suppose the agent observes a variable $a = a_0$ and would like to know the posterior distribution $P(b|a = a_0)$ for $b \in V$. This can be done by first updating the JPD to $P(V|a = a_0)$ and then marginalizing it to get $P(b|a = a_0)$.

To obtain $P(V|a = a_0)$, the product rule can be used:

$$P(V|a = a_0) = P(V, a = a_0)/P(a = a_0).$$

For $P(V, a = a_0)$, the observation $a = a_0$ forces $P(v) = 0$ for each $v \in D_V$, where $a \neq a_0$. Hence, these terms of $P(V)$ are set to zero. According to Eq. (2.7), the denominator $P(a = a_0)$ is just the sum of the remaining terms. Each remaining term is then divided by this sum. This operation is called *normalization*. Notice that the process involves updating $O(d^n)$ probability values. We refer to this difficulty as *updating intractability*.

Finally, marginalization yields

$$P(b|a = a_0) = \sum_{V \setminus \{b\}} P(V \setminus \{b\}, b|a - a_0).$$

This involves summing $O(d^n)$ probability values, which we refer to as *marginalization intractability*.

To illustrate the calculation (but not the intractability), consider a trivial example where $V = \{a, b\}$, $D_a = D_b = \{0, 1\}$, and $P(V)$ is given in Table 2.2. After setting terms inconsistent with $a = 0$ to zero in $P(a, b)$, Table 2.3 is obtained. After normalization, the intermediate result in Table 2.4 is obtained. After marginalization,

Table 2.2: *A JPD over* $V = \{a, b\}$

a	b	$P(a, b)$
0	0	0.1
0	1	0.2
1	0	0.3
1	1	0.4

Table 2.3: *Setting terms inconsistent*
with a = 0 to zero

a	b	$P(a, b|a = 0)P(a = 0)$
0	0	0.1
0	1	0.2
1	0	0
1	1	0

using a vector notation, we obtain the final result as follows:

$$P(b|a = 0) = \sum_a P(a, b|a = 0) = [p(b = 0|a = 0), p(b = 1|a = 0)]$$
$$= (1/3, 2/3).$$

We have seen that belief updating in large problem domains by using JPD directly suffers from acquisition intractability, updating intractability, and marginalization intractability. In the remaining sections of this chapter, we introduce graphical representations of probabilistic knowledge known as Bayesian networks. As we shall see, they allow effective acquisition and inference in many practical domains. The word *effective* is used loosely in this book to refer to a method or algorithm that is *efficient* on an average input instance but can be intractable in the worst case.

2.4 Graphs

The fundamental idea underlying effective representation and inference with probabilistic knowledge is that in the real world not every variable is *directly* dependent on every other variable. The output of a digital gate is directly dependent on its input and the state of the gate. Given the input and the state of the gate, the condition of the rest of the circuit is irrelevant to the output. This feature of the world is termed *conditional independence:* The output of the gate is independent of the rest of the

Table 2.4: *Posterior distribution*
$P(a, b|a = 0)$

a	b	$P(a, b|a = 0)$
0	0	1/3
0	1	2/3
1	0	0
1	1	0

circuit conditioned on the input and state of the gate. Belief updating using JPD is inefficient because it does not take advantage of the independence and is based on a very conservative assumption that everything is directly dependent on everything else.

To exploit the conditional independence, a graph can be used in which each node represents a variable or a subset of variables, and each edge (or link) signifies the direct dependence between the variables being connected. Two variables are conditionally independent if other variables block connections between them in the graph. The links in the graph may also be directed to signify an asymmetric dependence. Depending on whether we want links directed, we can use *direct graphs* in which all links are directed, *undirected graphs* in which all links are undirected, or *hybrid graphs* in which some links are directed. We will introduce the relevant graph-theoretic concepts throughout the book as they become necessary. Definitions of some basic concepts for graphs are given in the paragraphs that follow.

A *graph* is an ordered pair $G = (V, E)$. $V = (v_i | 1 \leq i \leq n, n > 0)$ denotes a set of nodes, where each v_i is called a *node* or a *vertex*. E is defined differently depending on whether G is a directed, undirected, or hybrid graph. If G is an undirected graph, then $E = (\langle u, v \rangle | u, v \in V, u \neq v)$, where each $\langle u, v \rangle$ is called a *link* or an *edge* and u and v are called *endpoints* of the link. The link $\langle u, v \rangle$ is said to be *incident* to the nodes u and v. Note that a link from a node to itself is disallowed because each link is intended to model the dependence between two distinct variables. The notations $\langle u, v \rangle$ and $\langle v, u \rangle$ denote the same link. Figure 2.4(a) shows an undirected graph with five nodes.

If G is a *directed graph*, then $E = [(u, v) | u, v \in V, u \neq v]$, where each (u, v) is called an *arc*. An arc (u, v) is directed from u (called the *tail* of the arc) to v (called the *head* of the arc). We also refer to u as a *parent* of v and to v as a *child* of u. When the direction of an arc (u, v) is not the current concern, we will treat it as a link $\langle u, v \rangle$. Figure 2.4(b) shows a directed graph.

If G is a *hybrid graph*, then E consists of both links and arcs. Figure 2.4(c) shows a hybrid graph.

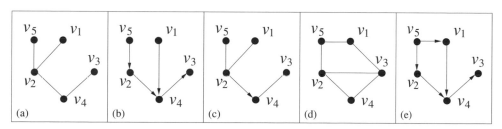

Figure 2.4: (a) A undirected graph. (b) A directed graph. (c) A hybrid graph. (d) A multiply connected undirected graph. (e) A multiply connected DAG.

For any node v, the *degree* $d(v)$ is the number of links incident to v. In a directed graph, the *in-degree* $d^-(v)$ is the number of arcs with head v, and the *out-degree* $d^+(v)$ is the number of arcs with tail v. A node v is a *root* if $d^-(v) = 0$. A node v is a *leaf* if $d^+(v) = 0$. If v is neither a root nor a leaf, then it is an *internal* node. In an undirected graph, a node v is a *terminal* node if $d(v) \leq 1$. Otherwise, v is *internal*. In Figure 2.4(a), $d(v_2) = 3$. The node v_1 is terminal and v_2 is internal. In Figure 2.4(b), $d^-(v_4) = 2$, $d^+(v_2) = 1$. The node v_1 is a root, v_3 is a leaf, and v_2 is internal.

Two nodes u and v are *adjacent* if $\langle u, v \rangle \in E$. A set of nodes adjacent to a node v (excluding v) is the *adjacency* of v, which is denoted by *adj*(v). A *path* is a sequence of nodes such that each pair of consecutive nodes is adjacent. A path is *simple* if no two nodes in the path are identical, except that the first node may be identical to the last node. We denote a path by enclosing the node sequence in $\langle \rangle$. In Figure 2.4(a), $\langle v_1, v_2, v_4, v_3 \rangle$ is a simple path. Note that $\langle v_1, v_2, v_4, v_3 \rangle$ and $\langle v_3, v_4, v_2, v_1 \rangle$ denote the same path. In this book, only simple paths are considered; hence, we drop the word *simple* whenever no confusion is possible. The *length* of a path with a sequence of $k \geq 2$ nodes is $k - 1$.

A path ρ in a directed graph is a *directed path* if each node in ρ, other than the first and the last, is the head of one arc in ρ and the tail of the other arc in ρ. If there is a directed path from a node u to a node v, then u is called an *ancestor* of v and v is called a *descendant* of u. We denote a directed path by enclosing the node sequence in (). In Figure 2.4(b), (v_5, v_2, v_4, v_3) is a directed path, and v_5 is an ancestor of v_3.

A path is a *cycle* if it contains two or more distinct nodes and the first node is identical to the last node. A cycle θ in a directed graph is *directed* if each node in θ is the head of one arc in θ and the tail of the other arc in θ. Otherwise, θ is *undirected*. We denote undirected cycles by enclosing the node sequence in $\langle \rangle$ and denote directed cycles using (). A directed graph is *acyclic* or is a directed acyclic graph if it contains no directed cycles. A directed acyclic graph is often referred to as a DAG. In Figure 2.4(d), $\langle v_1, v_5, v_2, v_3, v_1 \rangle$ is a cycle. The cycle in Figure 2.4(e) is undirected, and therefore the graph is a DAG.

A graph is *connected* if there exists a path between every pair of nodes. A connected graph is a *tree* if there exists exactly one path between every pair of nodes; otherwise, it is *multiply connected*. All graphs in Figure 2.4 are connected. The graphs in (a), (b), and (c) are trees. The graphs in (d) and (e) are multiply connected.

Given two graphs $G = (V, E)$ and $G' = (V', E')$, we say G' is a *subgraph* of G if $V' \subseteq V$ and $E' \subseteq E$, where endpoints of each link or arc in E' are elements of V'. The graph in (b) is a subgraph of (e), but (a) is not a subgraph of (d). A subgraph $G' = (V', E')$ of $G = (V, E)$ is *spanned* by a subset $V' \subset V$ of nodes if E' contains exactly those links of E with both endpoints in V'.

2.5 Bayesian Networks

Effective representation of probabilistic knowledge using graphical structures is based on conditional independence, as hinted in Section 2.4. We define the notion formally below.

Definition 2.1 *Let X, Y, and Z be disjoint subsets of variables of V. Subsets X and Y are* **conditionally independent** *given Z, denoted by I(X, Z, Y), if and only if for every x and for every y, z such that P(y, z) > 0, the following holds:*

$$P(x|y, z) = P(x|z).$$

When Z is empty, X and Y are said to be **marginally independent**, *which is denoted by I(X, Ø, Y).*

It can be proven easily that $I(X, Z, Y)$ holds if and only if $I(Y, Z, X)$ holds, namely,

$$I(X, Z, Y) \Longleftrightarrow I(Y, Z, X). \tag{2.8}$$

Consider a simple example of conditional independence. Bad habits in tooth cleaning increase the chances of cavity formation, which in turn causes toothache. The domain can be modeled by three variables:

$$\text{habit} \in \{\text{good, bad}\}, \quad \text{cavity} \in \{\text{yes, no}\}, \quad \text{toothache} \in \{\text{yes, no}\}.$$

The JPD from a particular population is given in Table 2.5. To test if $I(h, c, t)$ holds, compute $P(t|c, h) = P(h, c, t)/P(c, h)$. As the result in Table 2.6 shows, $I(h, c, t)$ does hold.

To derive Table 2.5, one has to assess seven probability values (the last is determined by the negation rule (Eq. (2.6)). However, by the product rule (Eq. (2.4)), it

Table 2.5: *JPD for P(h, c, t)*

h	c	t	$P(h, c, t)$
good	yes	yes	0.0595
good	yes	no	0.0105
good	no	yes	0.0315
good	no	no	0.5985
bad	yes	yes	0.204
bad	yes	no	0.036
bad	no	yes	0.003
bad	no	no	0.057

Table 2.6: $P(t|c, h)$ *and* $P(t|c)$

t	c	h	$P(t|c, h)$	t	c	$P(t|c)$
yes	yes	good	0.85	yes	yes	0.85
yes	yes	bad	0.85	yes	no	0.05
yes	no	good	0.05	no	yes	0.15
yes	no	bad	0.05	no	no	0.95
no	yes	good	0.15			
no	yes	bad	0.15			
no	no	good	0.95			
no	no	bad	0.95			

follows that

$$P(t, c, h) = P(t|c, h)P(c|h)p(h) = P(t|c)P(c|h)p(h).$$

Hence, the following five values completely define Table 2.5.

$$P(h = \text{good}) = 0.7$$
$$p(c = \text{yes}|h = \text{good}) = 0.1 \quad p(c = \text{yes}|h = \text{bad}) = 0.8$$
$$p(t = \text{yes}|c = \text{yes}) = 0.85 \quad p(t = \text{yes}|c = \text{no}) = 0.05.$$

This suggests a more efficient way to obtain Table 2.5: First, identify the conditional independence $I(h, c, t)$ and then assess the five values above. The conditional independence $I(h, c, t)$ then ensures that the five values uniquely define the JPD.

The identification of $I(h, c, t)$ is aided by the perception of causality: Given the direct causes, a variable is conditionally independent of all other variables excluding its own effects. The conditional independence $I(h, c, t)$ is thus identified by recognizing *cavity* as the direct cause of *toothache*.

A Bayesian network (BN) encodes conditional independence through a DAG.

Definition 2.2 *A **Bayesian network** is a triplet $(V, G, \mathcal{P})$. V is a set of variables, G is a connected DAG whose nodes correspond one-to-one to members of V such that each variable is conditionally independent of its nondescendants given its parents. Denote the parents of $v \in V$ in G by $\pi(v)$; $\mathcal{P}$ is a set of probability distributions:*

$$\mathcal{P} = \{P(v|\pi(v))|v \in V\}.$$

The DAG G is commonly referred to as the (dependence) *structure* of the BN. Figure 2.5 shows a BN for the cavity example. Each distribution $P(v|\pi(v))$ is considered to be associated with (or stored in) the node v. Because nodes in G map one-to-one to variables in V, we will use *node* and *variable* interchangeably when there is no confusion.

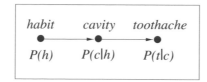

Figure 2.5: A Bayesian network for the cavity example.

Theorem 2.3 shows how the BN representation can help cope with the acquisition intractability. This theorem is commonly referred to as the *chain rule*.

Theorem 2.3 *Let* $(V, G, \mathcal{P})$ *be a Bayesian network. The following holds:*

$$P(V) = \prod_{v \in V} P(v|\pi(v)).$$

Proof: We prove by induction on $|V|$. The theorem is trivially true when $|V| = 1$. Assume it is true when $|V| = n - 1$.

Consider $|V| = n$. Because G is a DAG, it has at least one leaf a. Let $S = (V, G, \mathcal{P})$ and $S' = (V', G', \mathcal{P}')$ be obtained by removing a from S. By assumption, $P(V') = \prod_{v \in V'} P(v|\pi(v))$. By the product rule, $P(V) = P(a|V')P(V')$. Because S is a BN, by Definition 2.2, a is conditionally independent of its nondescendants given its parents. Hence, $P(a|V') = P(a|\pi(a))$. This implies $P(V) = P(a|\pi(a))P(V') = \prod_{v \in V} P(v|\pi(v))$. □

Consider the circuit in Figure 2.3 (duplicated in Figure 2.6(a)). The domain consists of 11 binary variables. A direct specification of the JPD must assess $2^{11} - 1 = 2047$ probability values. Figure 2.6(b) shows the DAG of a BN for this domain constructed by following the causal dependence among variables. For each of the seven root nodes, only one value is to be specified (e.g., $P(a = 1)$). For the

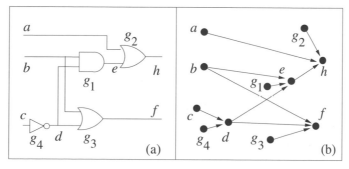

Figure 2.6: (a) A digital circuit. (b) A BN structure for the circuit.

output d of the NOT gate, four values are to be specified (e.g., $P(d = 1|c, g_4)$). For each output of the remaining gates, eight values are to be specified (e.g., $P(h = 1|a, e, g_2)$). Hence, a total of only 35 probability values needs to be specified.

2.6 Local Computation and Message Passing

We have seen that encoding conditional independence in the graphical structure of a BN can help cope with acquisition intractability. Can the graphical structure also help cope with the updating and marginalization intractability? The answer is yes. The basic idea is to avoid updating the JPD directly and then marginalizing down. Instead, we update local distributions that involve only a small number of variables and propagate their impact to other local distributions through message passing. The graphical dependence structure will aid us in deciding how small each local distribution should be and on which other distributions it might have an impact.

To illustrate this idea, consider the cavity example. Suppose that a child suffers from toothache and a diagnostic agent would like to update its belief about the child's habits in tooth-cleaning. In other words, the posterior distribution $P(h|t = y)$ is to be computed.

Recall from Figure 2.5 that the DAG has the topology $h \to c \to t$. The observation is at the node t, where $P(t|c)$ is stored. First, let the node t send a message along the arc between c and t:

$$P(t = y|c) = [0.85, 0.05].$$

Upon receiving the message, the node c computes its own message and sends it to the node h:

$$P(t = y|h) = \sum_c P(t = y, c|h) = \sum_c P(t = y|c, h)P(c|h)$$
$$= \sum_c P(t = y|c)P(c|h) = [0.13, 0.69],$$

where the third equality is due to $I(h, c, t)$. Note that the computation is based on the local distribution $P(c|h)$ and the incoming message $P(t = y|c)$. When h receives $P(t = y|h)$, it can compute

$$P(h|t = y) = \text{const } P(t = y|h)P(h) = [0.3054, 0.6946],$$

where const stands for a *constant*, which is determined through normalization.

In the preceding illustration, messages are sent by each node along the arcs of the DAG. The dimensions of distributions that each node x manipulates are no larger than that of $P(x|\pi(x))$. No direct manipulation of the JPD $P(h, c, t)$ is ever required. In fact, what we have shown are the operations performed by a general

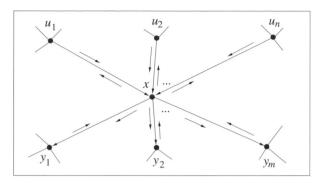

Figure 2.7: The $\lambda - \pi$ message passing at a typical node in a tree-structured BN.

algorithm called $\lambda - \pi$ *message passing* proposed by Pearl [52]. The algorithm is capable of computing in linear time the posterior distribution of every variable in a BN whose DAG is a tree. Figure 2.7 illustrates the message flow at a typical node during $\lambda - \pi$ message passing. Each arrow from x to a node v signifies a message sent from x to v that is obtained by combining all incoming messages into x except the one from v and the conditional probability distribution $P(x|\pi(x))$. After one message is passed along each direction of an arc, the posterior distribution of each variable can be obtained at the corresponding node.

Unfortunately, not all problem domains have tree-dependence structures. For example, the DAG in Figure 2.6(b) is multiply connected. Because the $\lambda - \pi$ message-passing algorithm is derived under the assumption of tree-dependence structures, there is no theoretical guarantee for the correctness of the posterior distributions when it is applied to multiply connected BNs. Experimental study (Murphy, Weiss, and Jordan [42]) applying $\lambda - \pi$ message passing to multiply connected BNs obtained posterior distributions highly correlated with the correct distributions in some BNs and oscillated posterior distributions in others.

2.7 Message Passing over Multiple Networks

One alternative method introduced by Pearl [52] to obtain correct posteriors from multiply connected BNs by $\lambda - \pi$ message passing is called *loop cutset conditioning*. We illustrate the method with an example. Consider a BN whose dependence structure is shown in Figure 2.8(a) in which each variable has a space $\{0, 1\}$. Suppose that $e = 1$ has been observed and we need to compute $P(c|e = 1)$.

To apply loop cutset conditioning, a node a is selected whose deletion would cut the cycle $\langle a, b, d, c, a \rangle$ open. For each value of a, a new BN is formed in which a is assumed to be observed at a particular value. Hence, two new BNs will be formed, as shown in Figure 2.8(b) and (c). Equivalently, the two new BNs can be

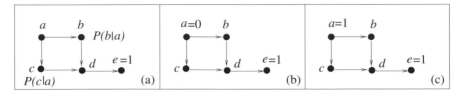

Figure 2.8: (a) The DAG structure of a BN.

represented, as in Figure 2.9. Note that a is now removed, and the observation of a is reflected in the distributions stored at nodes b and c. Because both BNs in Figure 2.9 are tree-structured, $\lambda - \pi$ message passing can be applied to each of them to obtain $P(c|a = 0, e = 1)$ (from (a)) and $P(c|a = 1, e = 1)$ (from (b)). To combine the two results and obtain $P(c|e = 1)$, the marginalization rule can be used:

$$P(c|e = 1) = P(c|a = 0, e = 1)P(a = 0|e = 1)$$
$$+ P(c|a = 1, e = 1)P(a = 1|e = 1),$$

where $P(a = 0|e = 1)$ and $P(a = 1|e = 1)$ remain to be computed. From Bayes's rule,

$$P(a = 0|e = 1) = P(e = 1|a = 0)P(a = 0)/P(e = 1),$$

and

$$P(a = 1|e = 1) = P(e = 1|a = 1)P(a = 1)/P(e = 1).$$

The expressions $P(a = 0)$ and $P(a = 1)$ can be obtained from the prior distribution $P(a)$, $P(e = 1|a = 0)$ can be obtained from $\lambda - \pi$ message passing in (a) without using the observation $e = 1$, and $P(e = 1|a = 1)$ can be obtained similarly in (b); $P(e = 1)$ is simply a normalizing constant.

We have seen how the hypothetical observation of a can be used to cut open the cycle and permit correct application of $\lambda - \pi$ message passing. In general, if the BN has k cycles, then $O(k)$ variables must be hypothetically observed to cut open all

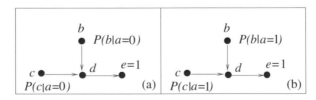

Figure 2.9: (a) The DAG structure of a BN.

cycles, and $\lambda - \pi$ message passing in $O(2^k)$ new BNs (one for each configuration of the $O(k)$ variables) needs to be performed. Therefore, as far as the behavior of message passing is concerned, we will refer to loop cutset conditioning as message passing *over multiple networks*. We refer to the standard $\lambda - \pi$ message passing in tree-structured BNs as message passing *over a single network*.

2.8 Approximation with Massive Message Passing

Another class of alternative methods uses stochastic simulation to compute approximate posteriors. Given a BN over a set V of variables and the observation $X = x$ of a subset X of variables, a stochastic simulation method computes the posterior distribution of each unobserved variable $v \in V$ as follows: It generates a sufficiently large number of configurations of V randomly from the JPD $P(V)$ defined by the BN. The posterior probability for $v = v_0$ is then approximated by

$$P(v = v_0 | X = x) \approx \frac{\text{Number of configurations where } X = x \text{ and } v = v_0}{\text{Number of configurations where } X = x} \quad (2.9)$$

As an example, consider the digital circuit in Figure 2.6 and its BN representation in (b). Suppose that we have observed $b = 1$ and $h = 0$ and need to compute $P(v|b = 0, h = 1)$ for each other variable v. One of the simplest stochastic simulation methods is *logic sampling* or *forward sampling* proposed by Henrion [24]. Configurations of V are generated one at a time. For each configuration, the value for each variable is determined by starting from the root nodes of the BN and proceeding along the direction of arcs.

Suppose that the first root selected is c, which has a distribution

$$[P(c = 0) = 0.3, P(c = 1) = 0.7].$$

It is viewed as defining two bins $bin(c = 0) : [0, 0.3]$ and $bin(c = 1) : (0.3, 1]$. To determine the value of c, a random number α is generated from a uniform distribution over $[0, 1]$. Say, for example, the value is $\alpha = 0.4$. Because it falls into the bin $bin(c = 1)$, the value of c is set to 1. Similarly, we can obtain the values for other root variables, say

$$(a = 0, b = 1, c = 1, g_1 = \text{norm}, g_2 = \text{norm}, g_3 = \text{norm}, g_4 = ab).$$

To determine the values for remaining variables, we start with a variable such that the values of its parents have been determined, say, d. Suppose that the conditional probability distribution $P(d|c, g_4)$ specifies

$$[P(d = 0|c = 1, g_4 = ab) = 0.6, P(d = 1|c = 1, g_4 = ab) = 0.4]$$

and $\alpha = 0.2$ is generated. The value of d is then set to 0. Repeating this process, suppose we obtain

$$(a = 0, b = 1, c = 1, d = 0, e = 0, f = 1, h = 0, g_1 = \text{norm},$$

$$g_2 = \text{norm}, g_3 = \text{norm}, g_4 = ab).$$

Because this configuration is compatible with the observation $b = 1$ and $h = 0$, it contributes 1 to the denominator in Eq. (2.9). When $P(a = 0|b = 1, h = 0)$ is estimated, this configuration also contributes 1 to the numerator of Eq. (2.9) but will contribute 0 if $P(a = 1|b = 1, h = 0)$ is estimated because it is incompatible with $a = 1$.

On the other hand, if a configuration is generated as

$$(a = 0, b = 1, c = 1, d = 1, e = 1, f = 1, h = 1, g_1 = \text{norm},$$

$$g_2 = \text{norm}, g_3 = \text{norm}, g_4 = ab),$$

where the values for d, e, h differ from the preceding configuration, it will contribute 0 to both the denominator and numerator of Eq. (2.9) because it is incompatible with the observation $h = 0$. After a large number of configurations have been generated, the posteriors of desired variables can be estimated using Eq. (2.9). The larger the number of simulations, the more accurate is the estimation.

Because forward sampling produces approximate posteriors, it is commonly referred to as an *approximate* inference method. The $\lambda - \pi$ message passing in tree-structured BNs and the loop cutset conditioning are referred to as *exact* inference methods, for they produce exact posteriors.

From the message-passing point of view, forward sampling passes messages from root nodes along the direction of arcs to internal nodes and then to leaf nodes. Each message from a parent node x to a child node y is the value of x simulated for the current configuration instead of a distribution, as in $\lambda - \pi$ message passing. The child node y is ready to simulate its own value as soon as it has received values from all its parents. Hence, a message is passed along each arc for every configuration simulated. This implies that, unlike $\lambda - \pi$ message passing in tree-structured BNs, where exactly two messages need to be passed along each arc, a *massive* volume of messages needs to be passed in order to get reasonably accurate posteriors. We will refer to the forward sampling as a method of *massive message passing* and to the $\lambda - \pi$ message passing in tree-structured BNs as a method of *concise message passing*.

In Section 2.6, we noted that $\lambda - \pi$ message passing is limited to tree-structured BNs. Although it can be applied to multiply connected BNs through loop cutset conditioning, when the BN contains many cycles, $\lambda - \pi$ message passing over an exponential number of networks needs to be performed. Forward sampling, on the other hand, does not suffer from the existence of cycles. Instead, it is limited from a different perspective. We have shown that when a simulated configuration is incompatible with the observation, it contributes nothing to the estimation of the posteriors; the computation spent in simulating this configuration is thus wasted. When the observation corresponds to a rare event, the problem becomes particularly serious because most configurations are incompatible. Several alternative stochastic simulation methods exist (e.g., Geman and Geman [19]; Pearl [51]; Shachter and Poet [61]; Fung and Favero [18]; Jensen, Kong, and Kjarulff [27]; and Ortiz and Kaelbling [46]) with different degrees of improvement in efficiency.

The limitations of loop cutset conditioning as well as forward sampling are not unusual. It has been shown that probabilistic inference in multiply connected BNs is NP-hard in general no matter which exact inference methods (Cooper [8]) or approximate inference methods (Dagum and Luby [10]) are used. Therefore, it is unlikely that efficient inference algorithms can be developed for general multiply connected BNs. Instead, any algorithm is likely to perform well only for BNs with certain topological or distributional properties. In Chapters 3 through 5, we present an alternative *exact* method for inference in *multiply connected* BNs that passes *concise* messages in a *single* network. The method uses the so-called junction tree representation. Although all the preceding methods work well (subject to the worst-case limitations) for single-agent inference, in the second half of the book, we show that the junction tree method can be extended to inference in multiagent systems, whereas loop cutset conditioning and stochastic simulation do not seem to extend well into multiagent systems.

2.9 Bibliographical Notes

Intractabilities of belief updating by JPD were analyzed by Szolovits and Pauker [72]. Pearl [48] introduced Bayesian networks to probabilistic inference in intelligent systems. Since the publication of Pearl's 1988 book [52], several other books on probabilistic reasoning using graphical models have appeared, including Neapolitan [43]; Lauritzen [36]; Jensen [29]; Shafer [62]; Castillo et al. [6]; and Cowell [9]. Jensen [29] contains an extensive account for construction of Bayesian networks for many practical situations.

Definition 2.1 on conditional independence and the use of $I(,,)$ notation are consistent with the usage by Pearl [52]. Pearl originated an axiomatic system for the conditional independence relation $I(,,)$. Gaag and Meyer [76] proposed an enhancement to Pearl's axiomatic system for conditional independence when sets of variables may overlap. It has been shown (Studeny [70]) that no finite set of axioms can completely characterize the conditional independence relation.

The $\lambda - \pi$ message-passing algorithm was introduced by Pearl [48] and Kim and Pearl [33]. Loop cutset conditioning was proposed in Pearl [49]. Forward sampling was introduced by Henrion [24]. Improvements to these methods and alternative methods for probabilistic reasoning using graphical models are also reported in the proceedings of several annual or biannual international conferences such as the

- Conference on Uncertainty in Artificial Intelligence (UAI),
- International Conference on Information Processing and Management of Uncertainty (IPMU),
- Workshop on Artificial Intelligence and Statistics, and the
- International Joint Conference on Artificial Intelligence (IJCAI);

in the proceedings of many national and regional conferences (with international participation), including the

- National Conference on Artificial Intelligence (AAAI),
- Canadian Artificial Intelligence Conference (AI), and the
- Florida Artificial Intelligence Research Society Conference (FLAIRS);

and in journals such as

- *Approximate Reasoning*,
- *Networks*,
- *Artificial Intelligence*, and
- *Artificial Intelligence Research*.

2.10 Exercises

1. Prove the negation rule using axioms (Eqs. (2.1) through (2.4)).
2. Prove the marginalization rule using axioms (Eqs. (2.1) through (2.4)).
3. An AND gate with two uniformly distributed input signals produces incorrect output 80% of the time when it malfunctions. The gate has a 1% chance of failure. What is the posterior probability distribution $P(gate|input_1 = 0, output = 0)$? What is the posterior distribution $P(gate|input_1 = 0, output = 1)$?
4. Compute the JPD of the tooth-cleaning habit network by the chain rule.
5. Compute $P(h|t = \text{yes})$ using the JPD obtained in the last question.
6. Compute $P(t|h = \text{bad})$ by message passing in the network of Figure 2.5.
7. Contaminated oxygen as well as weakened health may cause hallucinations when a fighter pilot is on duty. An accident occurred recently caused by the pilot's hallucinating. Although the health report of the pilot is available, it was issued 3 months ago. Construct a BN for the investigator of the accident.
8. Use forward sampling to compute $P(h|t = \text{yes})$ from the tooth-cleaning habit network.
9. Assume that $obs = (a = 0, b = 1, c = 1, h = 1)$ has been observed for the digital circuit in Figure 2.6. Use forward sampling to compute $P(v|obs)$ for each unobserved variable v.

3

Belief Updating and Cluster Graphs

Chapter 2 introduced several methods using message passing as a mechanism for effective belief updating in BNs. The $\lambda - \pi$ message passing method along the arcs of a BN produces exact posteriors only in tree-structured BNs. Loop cutset conditioning requires converting a nontree BN into multiple tree-structured BNs and carrying out $\lambda - \pi$ message passing in each of them. The stochastic simulation can be applied directly to a nontree BN to compute approximate posteriors but requires massive message passing in the BN. In this chapter, we focus on *concise* message passing and will drop the word *concise* when there is no confusion. We explore the opportunities presented by reorganizing the DAG structure of a BN into a *cluster graph* structure. The objective is to develop an alternative exact method that uses concise message passing in a single cluster graph structure for belief updating with nontree BNs. A cluster graph consists of an interconnected set of clusters. Each cluster is a subset of nodes (variables) in the original BN. Message passing is performed between adjacent clusters in the cluster graph. We investigate under what conditions such message passing leads to correct belief updating.

Section 3.2 introduces cluster graphs. A set of conventions on how a cluster graph constrains message passing is outlined in Section 3.3. Section 3.4 addresses the relation between message passing in cluster graphs and $\lambda - \pi$ message passing in BNs and demonstrates that $\lambda - \pi$ message passing can be viewed as message passing in a particular cluster graph given a BN. Cycles in cluster graphs are studied in Sections 3.5 and 3.6 in relation to message passing. In particular, we study two classes of cycles in cluster graphs, *degenerate* and *nondegenerate* cycles, and analyze their impact on belief updating by message passing. Section 3.5 shows that belief updating cannot be achieved in a cluster graph with nondegenerate cycles, no matter how message passing is performed. Section 3.6 demonstrates that some degenerate cycles admit belief updating by message passing, whereas others do not. Because cluster graphs with only degenerate cycles are much less common, the results of these analyses formally establish the necessity of tree-structured cluster graphs for general message-based belief updating. Finally, Section 3.7 shows that only a subclass of cluster trees, called *junction trees*, can support belief updating with message passing. Reorganization of a BN into a junction tree will be presented in Chapter 4.

3.1 Guide to Chapter 3

In this chapter, the idea of belief updating by concise message passing in a single graphical structure is further explored. The $\lambda - \pi$ message passing does this when the graph is a tree-structured DAG. However, when the DAG is not a tree, $\lambda - \pi$ message passing directly in the DAG does not work. Because concise message passing in a single graphical structure is such a simple and attractive approach to belief updating, this chapter considers an alternative graph structure – cluster graphs.

Given a BN, a cluster graph can be constructed as follows: Group the variables of the BN so that each variable is contained in at least one group (may be in more than one) called a *cluster*. Draw a graph with each cluster as a node. If two clusters share some variables, they can be connected by a link called a *separator*, but they need not be. If two clusters share no variables, they cannot be connected. Label each cluster with its group of variables and label each separator with the variables shared by the two corresponding clusters. The resultant is a cluster graph. Figure 3.1 shows the DAG of a BN in (a) and a possible cluster graph in (b). Each cluster is drawn as an oval and labeled by its variables, and each separator is drawn as a link with a labeled box. For naming convenience, an extra name, Q_i, is given to each cluster as well. A formal definition of cluster graphs and more examples are presented in Section 3.2. You may wonder about the impact of the arcs in the BN on the construction of a cluster graph. That will be discussed in the next chapter.

If there is a single path between every pair of clusters, the cluster graph is a cluster tree (e.g., Figure 3.1(b)). Many (such as Spiegelhalter [67]; Lauritzen and Spiegelhalter [37]; Shafer, Shenoy, and Mellouli [64]; Jensen et al. [30]; and Madsen and Jensen [39]) have studied how to perform belief updating in cluster

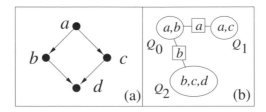

Figure 3.1: (a) A DAG. (b) A cluster graph.

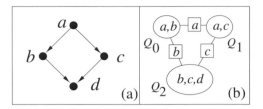

Figure 3.2: (a) A BN structure. (b) A cluster graph for (a).

trees. However, it is not known whether it is *impossible* to update belief by concise message passing in a nontree cluster graph. This chapter addresses this issue.

Section 3.3 specifies the rules of the game, namely, the rules for concise message passing in a cluster graph. Each cluster is associated with a probability distribution defined over its variables. Each cluster can only send messages to adjacent clusters. Each message must be a distribution defined over the corresponding separator variables. For example, the cluster Q_2 in Figure 3.1 can be associated with $P(b, c, d)$. It can send a message to the cluster Q_0, but it cannot send messages to Q_1 directly. The message that it sends to Q_0 can be $P(b)$ but cannot be $P(b, d)$.

Although cluster graphs appear to be quite different from DAGs, it turns out that $\lambda - \pi$ message passing in DAGs is just a special case of message passing in cluster graphs. Section 3.4 demonstrates that this is the case. It shows that $\lambda - \pi$ message passing corresponds to message passing in a particular cluster tree according to the rules outlined above. Hence, our study of message passing in cluster graphs unifies the study of all known concise message passing methods in a single graph structure.

Section 3.5 analyzes message passing using the cluster graph in Figure 3.2(b). Suppose that an initial belief state of the cluster graph is given, including distributions over Q_0, Q_1, and Q_2. The analysis demonstrates that infinitely many belief states of the cluster graph exist, each of which has the same distribution in cluster Q_0 as the given state, the same distribution in cluster Q_1, but distinct distribution in Q_2. Nevertheless, the messages sent from Q_2 to Q_0 are all identical as well as those sent to Q_1. That is, the messages are not sensitive to the infinitely many ways of difference on the initial distribution at Q_2. Hence, such a cluster graph cannot support belief updating by message passing in general no matter how the message passing is performed. The key feature of the cluster graph in (b) is its cycle in which no separator is contained in all other separators. We conclude that any cluster graphs with such cycles cannot support belief updating by message passing.

Section 3.6 considers cluster graphs with a different type of cycle in which one separator is contained in each other separator. Two such cluster graphs are shown in Figure 3.3. The analysis shows that belief updating by message passing is achievable in cluster graphs such as (a) but not in those such as (b). Furthermore, deleting a separator from either cluster graph does not change the situation. That is, belief

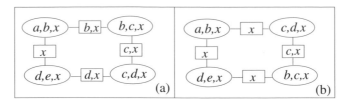

Figure 3.3: Cluster graphs with cycles in which one separator is contained in every other: (a) Message passing is achievable in this cluster graph, and (b) message passing cannot be performed correctly in this cluster graph.

updating can be achieved in a cluster tree obtained from (a) but not in a cluster tree obtained from (b).

Given a connected graph, a connected subgraph with the same nodes but the minimum number of links is a tree. That is, trees are the simplest subgraphs that retain connectedness. This simplicity makes them easier to analyze and more effi- cient to use. A cluster tree is more efficient to use than the corresponding cluster graph with cycles, because at least there are less separators to process.

This leads us to focus on cluster trees as the cluster graph structure for supporting belief updating with concise message passing. Section 3.7 demonstrates why some cluster trees such as the one in Figure 3.4 do not support belief updating by message passing. It is shown that a cluster tree can support belief updating with message passing if and only if it belongs to a subclass of cluster trees called *junction trees*. All known algorithms for belief updating by message passing use junction trees explic- itly or implicitly (e.g., $\lambda - \pi$ message passing). We now understand why this is so.

3.2 Cluster Graphs

Although concise message passing along the arcs of a BN (such as $\lambda - \pi$ mes- sage passing) does not yield exact posterior probability distributions in multiply connected BNs as discussed in Section 2.6, it is still possible to convert the DAG into a different structure so that local computation and concise message passing will achieve correct belief updating. We explore this possibility using a graph structure called a *cluster graph*. Unlike the directed or undirected graphs in which nodes are connected by arcs or links, a cluster graph consists of *clusters* that are

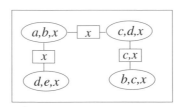

Figure 3.4: A cluster tree that cannot support belief updating by message passing.

interconnected. Each cluster is a set, and a connection between two clusters is labeled by the intersection between the two sets. In the DAG of a BN, a node represents a variable in a problem domain. In a cluster graph, each cluster denotes a subset of domain variables. In the DAG of a BN, an arc represents the direct causal dependence between two variables. In a cluster graph, a connection between two clusters signifies the direct probabilistic dependence between the two subsets of variables. The relevant graph-theoretic concepts are introduced in this section.

Given a set V, let $POW(V)$ denote its power set. The concept *junction graph* is defined as follows.

Definition 3.1 *A **junction graph** is a triplet (V, Ω, E). Set V is nonempty and is called the **generating set**, and Ω is a subset of $POW(V)$ such that $\cup_{Q \in \Omega} = V$. Each element Q of Ω is called a **cluster**.*

$$E = \{\langle Q_1, Q_2 \rangle | Q_1, Q_2 \in \Omega, Q_1 \neq Q_2, Q_1 \cap Q_2 \neq \emptyset\},$$

*where each unordered pair $\langle Q_1, Q_2 \rangle$ is called a **separator** between the two clusters Q_1 and Q_2 and is labeled by the intersection $Q_1 \cap Q_2$.*

A cluster is analogous to a node in an undirected graph, and a separator is analogous to a link. When there is no confusion, we will refer to a separator $\langle Q_1, Q_2 \rangle$ and the corresponding cluster intersection $Q_1 \cap Q_2$ interchangeably.

Figure 3.5 shows three junction graphs. Each cluster is shown as an oval, and each separator is shown as a box. For the junction graph in (c),

$$V = \{a, b, c, d\},$$
$$\Omega = \{Q_1 = \{a, b\}, Q_2 = \{a, c\}, Q_3 = \{b, c, d\}\},$$
$$E = \{\langle Q_1, Q_2 \rangle, \langle Q_1, Q_3 \rangle, \langle Q_2, Q_3 \rangle\}.$$

For simplicity, we often omit the separators in a figure. For instance, Figure 3.6(b) depicts the same junction graph as the one in Figure 3.5(b).

The concepts of *adjacent* clusters, *path*, and *cycle* in junction graphs have definitions similar to those of the same concepts applied to undirected graphs. For example, in Figure 3.5(b), the clusters $\{a, b\}$ and $\{a, e\}$ are adjacent. There are two paths

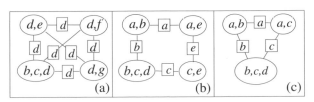

Figure 3.5: Junction graphs with clusters shown in ovals and separators shown in boxes.

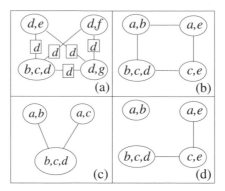

Figure 3.6: Cluster graphs.

between them: the path $\langle\{a, b\}, \{a, e\}\rangle$ and the path $\langle\{a, b\}, \{b, c, d\}, \{c, e\}, \{a, e\}\rangle$. In other words, there is a cycle in (b).

Let $H = (V, \Omega, E)$ be a junction graph. If some separators are removed from H, it is no longer a junction graph. The resultant $H' = (V, \Omega, E')$ is referred to as a *cluster graph* over V, where $E' \subset E$ is the remaining subset of separators. In general, a junction graph is a cluster graph (with no separators removed), but the reverse is not necessarily true. For example, Figure 3.5(a) is a junction graph as well as a cluster graph, whereas Figure 3.6(a) is a cluster graph but not a junction graph.

In a cluster graph H', if there exists exactly one path between each pair of clusters, H' is a *cluster tree* over V. If there exist more than one path between a pair of clusters, H' is *multiply connected*. If there exists no path between at least one pair of clusters, H' is *disconnected*. For example, Figure 3.6(c) is a cluster tree, (a) is a multiply connected cluster graph, and (d) is a disconnected cluster graph.

Let ρ be a cycle in a cluster graph. If there exists a separator S on ρ that is contained in every other separator, then ρ is a *degenerate cycle*. Otherwise, ρ is a *nondegenerate cycle*. All cycles in Figure 3.5(a) and the cycle in Figure 3.7(a) are degenerate because a separator $\{d\}$ is contained in every other separator. The cycles in Figure 3.5(b) and (c) and the cycle in Figure 3.7(b) are nondegenerate because no separator is contained in all other separators.

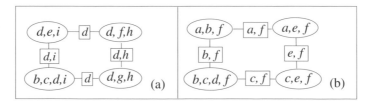

Figure 3.7: Cluster graphs with cycles.

A degenerate cycle ρ is a *strong* degenerate cycle if all separators in ρ are identical. Otherwise, ρ is a *weak* degenerate cycle. The cycles in Figure 3.5(a) are all strong degenerate cycles, whereas the cycle in Figure 3.7(a) is a weak degenerate cycle.

A nondegenerate cycle ρ is a *strong* nondegenerate cycle if $\cap_i S_i = \phi$, where i is over every separator S_i in ρ. Otherwise, ρ is a *weak* nondegenerate cycle. The cycle in Figure 3.5(c) is a strong nondegenerate cycle, whereas the cycle in Figure 3.7(b) is a weak nondegenerate cycle because $\cap_i S_i = \{f\}$.

As will be seen in the next section, degenerate cycles behave very differently from nondegenerate cycles as far as message passing in a cluster graph is concerned. The study of these cycles will lead to a subclass of cluster graphs that can support belief updating by message passing.

3.3 Conventions for Message Passing in Cluster Graphs

If the cluster graph is to be used to structure the representation of probabilistic knowledge and message passing, the representation and the content of the message must respect the structure of the cluster graph. The following conventions, which dictate how the structure of a cluster graph constrains the representation and message passing, are imposed:

1. The generating set of the cluster graph is the set V of domain variables. That is, each cluster must be a subset of domain variables, and the union of all clusters must cover the entire domain.
2. Each cluster Q (a subset of V) is associated with one or more nonnormalized (not summing to 1) probability distributions called *potentials* defined over Q or its subsets. A potential is equivalent to a probability distribution because it differs only by a normalizing constant and the constant can be removed at any time. We denote the potential over a set X of variables by $B(X)$. Using potentials instead of probability distributions affords the flexibility of not having to normalize out the constants for intermediate results during a long sequence of operations on some probability distributions. When the final result is obtained, one normalization is sufficient to acquire the intended probability distribution. As a simple example, consider the domain $V = \{a, b\}$, $D_a = D_b = \{0, 1\}$ with $P(a, b)$ given in Table 3.1. It was used in Section 2.3 to illustrate belief updating by JPD. Using the potential representation, when $a = 0$ is observed, the terms inconsistent with the observation are set to zero. The intermediate result is shown in Table 3.1 as the potential $B(a, b)$. To obtain the posterior probability distribution over b, the variable a is marginalized out of $B(a, b)$. The intermediate result is shown as the potential $B(b)$. Finally, $B(b)$ is normalized to obtain the end result $P(b|a = 0)$.

   A cluster graph with potentials attached is referred to as a *cluster graph representation* of the domain or simply a *cluster graph*.

Table 3.1: *Illustration of belief updating with potential representation*

a	b	$P(a, b)$	a	b	$B(a, b)$	b	$B(b)$	b	$P(b\|a = 0)$
0	0	0.1	0	0	0.1	0	0.1	0	1/3
0	1	0.2	0	1	0.2	1	0.2	1	2/3
1	0	0.3	1	0	0				
1	1	0.4	1	1	0				

3. A message can only be sent to an adjacent cluster, and the content of the message must be a potential over the corresponding separator. For example, if a problem domain is represented as the cluster graph in Figure 3.6(b), messages from the cluster $\{a, b\}$ can only be sent to the clusters $\{a, e\}$ and $\{b, c, d\}$ but not to the cluster $\{c, e\}$. The message from $\{a, b\}$ to $\{a, e\}$ must be a potential $B(a)$, and the message from $\{a, b\}$ to $\{b, c, d\}$ must be a potential $B(b)$.

In the remaining sections, we consider what is (or is not) achievable under these conventions.

3.4 Relation with $\lambda - \pi$ Message Passing

Message passing in a cluster graph can be viewed as a more general operation than $\lambda - \pi$ message passing in a BN. This is because the domain of the BN can be organized into many different cluster graphs with at least one of them containing the same set of message paths used in $\lambda - \pi$ message passing, as we demonstrate herein.

Consider the cluster graph in Figure 3.8 for the cavity example. Each cluster is made of $\{v\} \cup \pi(v)$ for a node v in Figure 2.5. The probability distribution stored in v in the BN is shown here as the potential associated with the cluster. Each message sent in Section 2.6 is shown as a potential over a separator.

The BN structure in the cavity example is a chain. The $\lambda - \pi$ algorithm is applicable to belief updating in any tree-structured BNs. To demonstrate that for any such BN a cluster graph exists that can support $\lambda - \pi$ message passing, we first briefly describe the basic scheme of the $\lambda - \pi$ algorithm. Figure 3.9 illustrates the

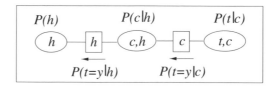

Figure 3.8: A cluster graph corresponding to message passing in the cavity BN.

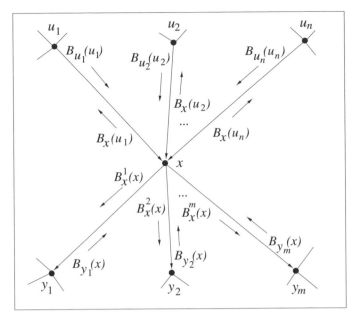

Figure 3.9: The $\lambda - \pi$ message passing at a typical node in a tree-structured BN.

messages passed in and out of a typical node in a BN according to the algorithm. Each message is denoted as a potential over a given variable sent from a given node. For instance, $B_x(u_1)$ is a potential over the variable u_1 sent from the node x. If the same node sends messages over the same variable to different destinations, we use superscripts to differentiate these messages. For example, in Figure 3.9, $B_x^2(x)$ is a potential over x sent from the node x to y_2. The source and destination of each message are indicated by the corresponding arrow in Figure 3.9. The exact form of each message and its formal derivation are not the focus here and can be found in Pearl [52]. In his formulation, the messages along the direction of arcs of the BN (e.g., $B_x^2(x)$) are called π messages, and those against the direction of arcs (e.g., $B_x(u_1)$) are called λ messages. Some key computational aspects of the algorithm are summarized as follows:

- Each node x in the BN is associated with a conditional probability distribution $P(x|\Pi(x))$, where $\Pi(x)$ is the set of parents of x. Note that the usual notation $\pi(x)$ has been replaced with $\Pi(x)$ to avoid confusion with the π message.
- Each message is a potential over a single variable. In particular, each λ message sent from x to a parent u is a potential over u, and each π message sent from x to a child y is a potential over x.
- Each message sent from x to an adjacent node z is computed from $P(x|\Pi(x))$ as well as all messages incoming to x except the message incoming from z.
- The posterior probability distribution of x is computed from $P(x|\Pi(x))$ and all messages incoming to x.

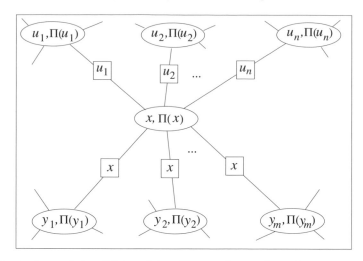

Figure 3.10: A cluster $\{x\} \cup \Pi(x)$ and its adjacent clusters in a cluster tree, where x is a typical node in a tree-structured BN with its parents $\Pi(x)$.

Given a tree-structured BN, a cluster graph H is created as follows: For each node x with its parents $\Pi(x)$ in the BN, create a cluster $\{x\} \cup \Pi(x)$ in H. If u is a parent of x in the BN, connect the clusters $\{u\} \cup \Pi(u)$ and $\{x\} \cup \Pi(x)$ in H. Because $u \in \Pi(x)$ and the BN is tree-structured, the separator between the two clusters is $\{u\}$. Figure 3.10 shows the cluster $\{x\} \cup \Pi(x)$ and its adjacent clusters in H. Some key topological features of H are summarized below:

- There is a one-to-one mapping between the clusters in H and the nodes in the BN. In particular, for each node x with its parents $\Pi(x)$ in the BN, there is a unique cluster $\{x\} \cup \Pi(x)$ in H.
- If x has n parents and m children in the BN, the cluster $\{x\} \cup \Pi(x)$ has exactly $n + m$ adjacent clusters in H.
- H is a cluster tree.
- For each parent u of x in the BN, there is a cluster $\{u\} \cup \Pi(u)$ in H adjacent to the cluster $\{x\} \cup \Pi(x)$ such that their separator is $\{u\}$.
- For each child y of x in the BN, there is a cluster $\{y\} \cup \Pi(y)$ in H adjacent to the cluster $\{x\} \cup \Pi(x)$ such that their separator is $\{x\}$.

The cluster tree H can support $\lambda - \pi$ message passing as follows:

- For each λ message sent from x to a parent u in the BN, the separator between clusters $\{x\} \cup \Pi(x)$ and $\{u\} \cup \Pi(u)$ in H is just the variable u over which the message is defined. Hence, the λ message can be sent along the separator according to our convention (Section 3.3).
- For each π message sent from x to a child y in the BN, the separator between clusters $\{x\} \cup \Pi(x)$ and $\{y\} \cup \Pi(y)$ in H is just the variable x over which the message is defined.

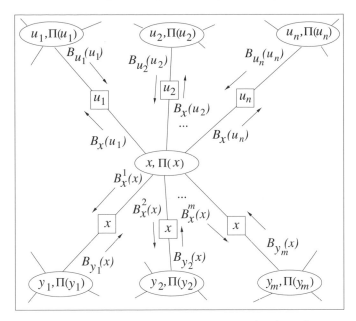

Figure 3.11: Illustration of the $\lambda - \pi$ message passing in a cluster graph.

- For each cluster $\{x\} \cup \Pi(x)$, the distribution $P(x|\Pi(x))$ can be associated with it. Hence, the cluster has all the information necessary to process the incoming and outgoing λ and π messages and to compute the posterior distribution of x.

Figure 3.11 illustrates how each message in Figure 3.9 can be passed in the cluster tree H. From the preceding comparison, we conclude that $\lambda - \pi$ message passing corresponds to message passing in a particular type of cluster tree.

Because $\lambda - \pi$ message passing cannot produce correct posterior distributions for multiply connected BNs in general, we investigate in the remaining sections what types of cluster graph structures allow such computation. It will be seen that multiply connected cluster graphs do not support belief updating by message passing in general.

3.5 Message Passing in Nondegenerate Cycles

We investigate message passing in a cluster graph with nondegenerate cycles. We start with a domain that has the dependence structure of Figure 3.12(a). All variables are assumed binary. The variables can be organized into the cluster graph in (b). That is, the generating set is $V = \{a, b, c, d\}$, and there are three clusters $Q_0 = \{a, b\}$, $Q_1 = \{a, c\}$, and $Q_2 = \{b, c, d\}$. Note that there is a strong nondegenerate cycle in (b).

48 *Belief Updating and Cluster Graphs*

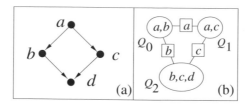

Figure 3.12: (a) A DAG dependence structure. (b) A cluster graph of (a) with a nondegenerate cycle.

Given a cluster Q, let $B_Q(Q)$ signify the potential associated with Q, where the subscript denotes the associated cluster, and the argument inside "()" represents the domain of the potential. When there is no confusion, the subscript will be dropped.

Each cluster in (b) is associated with a potential $B(a, b) = P(a, b)$, $B(a, c) = P(a, c)$, and $B(b, c, d) = P(b, c, d)$, respectively. It is assumed that the potentials satisfy

$$\sum_b B(a, b) = \sum_c B(a, c), \quad \sum_a B(a, b) = \sum_{c,d} B(b, c, d),$$

$$\sum_a B(a, c) = \sum_{b,d} B(b, c, d),$$

and H is said to be *locally consistent*. In general, if two adjacent clusters Q and Q' satisfy

$$\sum_{Q \setminus Q'} B(Q) = \text{const} \sum_{Q' \setminus Q} B(Q'),$$

where const stands for a positive constant, then Q and Q' are said to be *consistent*. If every pair of adjacent clusters is consistent, the cluster graph is *locally consistent*. The cluster graph representation is denoted by H. Because H is locally consistent, if any cluster Q with an adjacent cluster Q' passes a potential over their separator, the message cannot change $B(Q')$.

Suppose that $d = d_0$ is observed and each potential is to be updated to the corresponding posterior:

$$B(a, b) \to P(a, b | d = d_0), \quad B(a, c) \to P(a, c | d = d_0),$$
$$B(b, c, d) \to P(b, c, d | d = d_0).$$

For Q_2, the method in Section 2.3 (treating $B(b, c, d)$ as a JPD) can be used to obtain $P(b, c, d | d = d_0)$ locally. For Q_0 and Q_1, belief updating must rely on the message $P(b | d = d_0)$ sent by Q_2 to Q_0 and the message $P(c | d = d_0)$ sent by Q_2 to Q_1. In the following discussion, we show that in general it is impossible for Q_0 and Q_1 to update their potentials correctly based on these messages.

Before presenting the general result, we illustrate with a particular JPD over V. From Theorem 2.3, we can independently specify $P(a)$, $P(b|a)$, $P(c|a)$, and $P(d|b, c)$. A JPD is thus specified as follows, where, for simplicity, $P(b = b_0|a = a_0)$ is written as $P(b_0|a_0)$:

$$P(a_0) = 0.26$$
$$P(b_0|a_0) = 0.98 \qquad P(b_0|a_1) = 0.33$$
$$P(c_0|a_0) = 0.02 \qquad P(c_0|a_1) = 0.67$$
$$P(d_0|b_0, c_0) = 0.03 \quad P(d_0|b_0, c_1) = 0.66$$
$$P(d_0|b_1, c_0) = 0.7 \qquad P(d_0|b_1, c_1) = 0.25$$

Denote the JPD defined according to the preceding distributions by *jpd*. The cluster potentials are assigned as follows:

$$B(a, b) = P(a)P(b|a), \quad B(a, c) = P(a)P(c|a), \quad B(b, c, d) = P(b, c)P(d|b, c),$$

where $P(b, c) = \sum_a P(a)P(b|a)P(c|a)$. This initial state of H is referred to as s and is detailed below:

$B(a, b)$:	$P(a_0, b_0) = 0.2548$		$P(a_0, b_1) = 0.0052$
	$P(a_1, b_0) = 0.2442$		$P(a_1, b_1) = 0.4958$
$B(a, c)$:	$P(a_0, c_0) = 0.0052$		$P(a_0, c_1) = 0.2548$
	$P(a_1, c_0) = 0.4958$		$P(a_1, c_1) = 0.2442$
$B(b, c, d)$:	$P(b_0, c_0, d_0) = 0.0050613$		$P(b_0, c_0, d_1) = 0.16364871$
	$P(b_0, c_1, d_0) = 0.21799143$		$P(b_0, c_1, d_1) = 0.11229861$
	$P(b_1, c_0, d_0) = 0.23260301$		$P(b_1, c_0, d_1) = 0.09968701$
	$P(b_1, c_1, d_0) = 0.042177506$		$P(b_1, c_1, d_1) = 0.12653252.$

Clearly, H is locally consistent under s. Suppose $d = d_0$ is then observed. The message from Q_2 to Q_0 is computed using the method in Section 2.3, and $P(b|d_0) = (0.448, 0.552)$ is obtained. Similarly, the message from Q_2 to Q_1 is $P(c|d_0) = (0.477, 0.523)$.

Now consider a different JPD that differs from *jpd* by replacing $P(d|b, c)$ with the following:

$$P'(d_0|b_0, c_0) = 0.533604 \quad P'(d_0|b_0, c_1) = 0.115431$$
$$P'(d_0|b_1, c_0) = 0.14 \qquad P'(d_0|b_1, c_1) = 0.66.$$

Denote this JPD by *jpd'*. If cluster potentials are assigned accordingly, $B'(b, c, d) \neq B(b, c, d)$ but $B'(a, b) = B(a, b)$ and $B'(a, c) = B(a, c)$. This initial state of H is referred to as s' and H is locally consistent at s'. Now after $d = d_0$ is observed, if the messages $P'(b|d_0)$ and $P'(c|d_0)$ are computed, they will be identical to those obtained from state s (see Exercise 3). That is, the messages are insensitive to the difference between the two initial states. Because the initial states are

$$B'(a, b) = B(a, b), \quad B'(a, c) = B(a, c),$$

and messages are

$$P'(b|d_0) = P(b|d_0), \quad P'(c|d_0) = P(c|d_0),$$

the posterior distributions in Q_0 and Q_1 will be identical in the two cases.

One may ask whether this should be the case. To find out, the chain rule can be used to obtain $P(a, b, c, d)$ and $P'(a, b, c, d)$, and then the belief updating method in Section 2.3 can be applied. From s, $P(a_1|d_0) = 0.666$ is obtained, and from s', $P'(a_1|d_0) = 0.878$ is obtained (see Exercise 4). The difference is significant.

We now show that the phenomenon above is not accidental. Without losing generality, it is assumed that the JPDs involved are strictly positive. Lemma 3.2 says that, for infinitely many different initial potentials of Q_2, the messages from Q_2 to Q_0 (Q_1) are identical.

Lemma 3.2 *Let jpd be a strictly positive JPD over V and s be a locally consistent state of H derived from jpd.*

Let jpd' be another JPD identical to jpd in $P(a)$, $P(b|a)$, and $P(c|a)$ but distinct in $P(d|b, c)$. Let s' be the locally consistent state of H from jpd'. The jpd' and s' are such that the message $P(b|d = d_0)$ ($P(c|d = d_0)$) produced from s' is identical to that produced from s.

Then, given jpd, the number of distinct jpd's potentials that satisfy the above conditions is infinite.

Proof: The message component $P(b_0|d_0)$ can be expanded as

$$P(b_0|d_0) = P(b_0, d_0)/(P(b_0, d_0) + P(b_1, d_0)) = \left[1 + \frac{P(b_1, d_0)}{P(b_0, d_0)}\right]^{-1}$$

$$= \left[1 + \frac{P(b_1, c_0, d_0) + P(b_1, c_1, d_0)}{P(b_0, c_0, d_0) + P(b_0, c_1, d_0)}\right]^{-1}$$

$$= \left[1 + \frac{P(d_0|b_1, c_0)P(b_1, c_0) + P(d_0|b_1, c_1)P(b_1, c_1)}{P(d_0|b_0, c_0)P(b_0, c_0) + P(d_0|b_0, c_1)P(b_0, c_1)}\right]^{-1}.$$

Similarly, the message component $P(c_0|d_0)$ can be expanded as

$$P(c_0|d_0) = \left[1 + \frac{P(c_1, d_0)}{P(c_0, d_0)}\right]^{-1}$$

$$= \left[1 + \frac{P(d_0|b_0, c_1)P(b_0, c_1) + P(d_0|b_1, c_1)P(b_1, c_1)}{P(d_0|b_0, c_0)P(b_0, c_0) + P(d_0|b_1, c_0)P(b_1, c_0)}\right]^{-1}.$$

According to the assumption, s and s' agree on $P(a, b)$, $P(a, c)$, and $P(b, c)$ but differ in $P(d|b, c)$. If messages from s' are identical to that from s, namely, $P'(b|d_0) = P(b|d_0)$ and $P'(c|d_0) = P(c|d_0)$, then $P'(d|b, c)$ must be the solutions of the following equations:

$$\frac{P'(d_0|b_1, c_0)P(b_1, c_0) + P'(d_0|b_1, c_1)P(b_1, c_1)}{P'(d_0|b_0, c_0)P(b_0, c_0) + P'(d_0|b_0, c_1)P(b_0, c_1)} = \frac{P(b_1, d_0)}{P(b_0, d_0)}$$

$$\frac{P'(d_0|b_0, c_1)P(b_0, c_1) + P'(d_0|b_1, c_1)P(b_1, c_1)}{P'(d_0|b_0, c_0)P(b_0, c_0) + P'(d_0|b_1, c_0)P(b_1, c_0)} = \frac{P(c_1, d_0)}{P(c_0, d_0)}.$$

Because $P'(d|b, c)$ has four independent parameters but is constrained by only two equations, it has *infinitely* many solutions. Each solution defines an initial state s' of H that satisfies all conditions in the lemma. □

Lemma 3.3 says that, with such difference in initial states as specified in Lemma 3.2, correct belief updating will produce distinct posteriors.

Lemma 3.3 *Let jpd be a strictly positive JPD over V and jpd' be another JPD identical to jpd in $P(a)$, $P(b|a)$, and $P(c|a)$ but distinct in $P(d|b, c)$.*

Then $P(a|d = d_0)$ produced from jpd' is distinct from that produced from jpd in general.

Proof: From *jpd*, it follows

$$P(a|d_0) = \sum_{b,c} P(a|b, c)P(b, c|d_0). \tag{3.1}$$

From *jpd'*, it follows

$$P'(a|d_0) = \sum_{b,c} P(a|b, c)P'(b, c|d_0), \tag{3.2}$$

where $P(a|b, c)$ has been used because *jpd'* is identical with *jpd* in $P(a)$, $P(b|a)$, and $P(c|a)$. If $P(b, c|d_0) \neq P'(b, c|d_0)$ (which is shown below), then in general $P(a|d_0) \neq P'(a|d_0)$.

The following hold:

$$P(b, c|d_0) = \frac{P(d_0|b, c)P(b, c)}{P(d_0)} = \frac{P(d_0|b, c)P(b, c)}{\sum_{b,c} P(d_0|b, c)P(b, c)},$$

$$P'(b, c|d_0) = \frac{P'(d_0|b, c)P(b, c)}{P'(d_0)} = \frac{P'(d_0|b, c)P(b, c)}{\sum_{b,c} P'(d_0|b, c)P(b, c)}.$$

Because $P(d|b, c) \neq P'(d|b, c)$ by assumption, in general $P(b, c|d_0) \neq P'(b, c|d_0)$. $\square$

Theorem 3.4 *Belief updating cannot be achieved using the cluster graph H in general no matter how message passing is performed.*

Proof: By Lemma 3.2, messages (e.g., $P(b|d = d_0)$ and $P(c|d = d_0)$) from H are insensitive to the initial states of H, and hence the posterior distributions (e.g., $P(a|d = d_0)$) computed based on these messages cannot be sensitive to the initial states either. However, by Lemma 3.3, the posterior distributions should be different in general given different initial states. Hence, correct belief updating cannot be achieved in H. $\square$

Note that the preceding difficulty is due to the nondegenerate cycle in H. Observe in Eqs. (3.1) and (3.2) that correct belief updating requires $P(b, c|d_0)$. According to the message-passing conventions (Section 3.3), to pass such a message, a separator must contain $\{b, c\}$, the intersection between Q_2 and $Q_0 \cup Q_1$. The existence of the nondegenerate cycle signifies the splitting of such a separator (into separators $\{b\}$ and $\{c\}$). The result is the passing of marginals of $P(b, c|d_0)$ (the insensitive messages) and ultimately incorrect belief updating.

This analysis can be generalized to an arbitrary, strong nondegenerate cycle of length 3 (the length of the cycle in H), where each of a, b, c, d is a set of variables. The result in Lemmas 3.2, 3.3, and Theorem 3.4 can be similarly derived (Exercise 5).

This analysis can be further generalized to an arbitrary, strong nondegenerate cycle of length $K > 3$. By clumping $K - 2$ adjacent clusters into one big cluster Q, the cycle is reduced to length 3. Any message passing among the $K - 2$ clusters can be considered as occurring in the same way as before the clumping but "inside" Q. Now the analysis above for an arbitrary, strong nondegenerate cycle of length 3 applies.

Furthermore, the result can be generalized to an arbitrary, weak nondegenerate cycle ρ of length $K \geq 3$. Let

$$\rho = \langle Q_0, Q_1, \ldots, Q_{K-1}, Q_0 \rangle,$$

the separator between Q_i and Q_{i+1} be S_i ($0 \le i < K - 1$), the separator between Q_{K-1} and Q_0 be S_{K-1}, and $R = \cap_{i=0}^{K-1} S_i$. Let the potential of each cluster be

$$B_{Q_i}(Q_i) = \text{const } P_{Q_i}(Q_i \backslash R | R) P_{Q_i}(R).$$

Note that $P_{Q_i}(R)$ and $P_{Q_j}(R)$ may be different due to, say, an observation on $r \in R$ available to Q_i but not to Q_j. Message passing in ρ can be considered as independent passing of two message streams, one determined by $P_{Q_i}(Q_i \backslash R | R)$ and one determined by $P_{Q_i}(R)$. For example, the message from Q_0 to Q_1 is

$$B_{Q_0}(S_0) = \text{const } P_{Q_0}(S_0 \backslash R | R) P_{Q_0}(R),$$

where $P_{Q_0}(S_0 \backslash R | R)$ can be obtained by marginalization of $P_{Q_0}(Q_0 \backslash R | R)$.

The first message stream according to $P_{Q_i}(Q_i \backslash R | R)$ is equivalent to the message passing in a strong nondegenerate cycle

$$\rho' = \langle Q_0', Q_1', \dots, Q_{K-1}', Q_0' \rangle,$$

where each cluster $Q_i' = Q_i \backslash R$. According to the preceding analysis, belief updating cannot be achieved by message passing in ρ'. The second message stream according to $P_{Q_i}(R)$ is straightforward. It in general has no impact on the first message stream. Therefore, belief updating cannot be achieved by message passing in ρ in general.

To summarize, the difficulty will arise whenever a cluster graph contains nondegenerate cycles. This is stated in the following corollary.

Corollary 3.5 *In general, belief updating cannot be achieved in a cluster graph with nondegenerate cycles no matter how message passing is performed.*

3.6 Message Passing in Degenerate Cycles

Can a cluster graph with only degenerate cycles support belief updating? We consider the strong and weak degenerate cycles separately, for they behave differently. Consider a cluster graph with K clusters Q_i ($i = 0, 1, \dots, K - 1$) forming a strong degenerate cycle where every separator is X (see Figure 3.13). Suppose that

$$I(Q_i \backslash X, X, Q_j \backslash X)$$

holds, and each cluster potential is set to $B_{Q_i}(Q_i) = P(Q_i)$. If $a \in Q_0$ is observed to be a_0, we can update $B_{Q_0}(Q_0)$ to $P(Q_0 | a_0)$ and then send a message

$$P(X | a_0) = \sum_{Q_0 \backslash X} P(Q_0 | a_0)$$

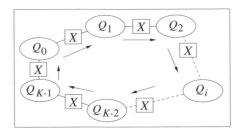

Figure 3.13: A cluster graph with a strong degenerate cycle.

to an adjacent cluster, say, Q_1, as shown by the arrow in Figure 3.13. The cluster potential $B_{Q_1}(Q_1)$ can then be updated to

$$P(Q_1|a_0) = P(Q_1 \backslash X|X)P(X|a_0) = P(Q_1)P(X|a_0)/P(X),$$

where both $P(Q_1)$ and $P(X)$ are locally available in Q_1. Message passing is repeated along the cycle until Q_0 receives what it sent and terminates the process. Hence, we conclude that a strong degenerate cycle *does* support belief updating.

Is the cyclic structure of a strong degenerate cycle necessary for belief updating? The answer is no. Clearly, belief updating can be performed if the cycle is broken at any separator into a cluster chain, as shown in Figure 3.14. The cluster Q_0 can then send the preceding message to adjacent cluster Q_1, which in turn passes on to the next adjacent cluster. The arrows in Figure 3.14 illustrate this process. When the message is received by the cluster Q_{K-1}, it has no one to send the message to and the process halts.

Next, consider a weak degenerate cycle ρ in which separators are not all identical but there exists a separator S that is contained in each other separator. Let the cycle be

$$\rho = \langle Q_0, Q_1, \ldots, Q_{K-1}, Q_0 \rangle,$$

where $K \geq 3$. Let the clusters connected by S be Q_0 and Q_1. There are two paths

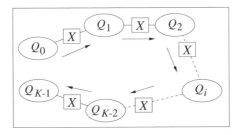

Figure 3.14: Breaking a strong degenerate cycle into a cluster chain.

between Q_0 and Q_1,

$$\langle Q_0, Q_1 \rangle \quad \text{and} \quad \langle Q_0, Q_{K-1}, Q_{K-2}, \ldots, Q_2, Q_1 \rangle.$$

The message that can be passed from Q_0 to Q_1 along S is a potential

$$B_{Q_0}(S) = \text{const } P_{Q_0}(S).$$

Because the message passed along any separator $S' \neq S$ can be expressed as a potential

$$B(S') = \text{const } P(S'\backslash S|S)P(S),$$

which contains $P(S)$, the path $\langle Q_0, Q_1 \rangle$ is *redundant:* the same information can be propagated through the other path. Therefore, whether or not belief updating is achievable by message passing in a weak degenerate cycle can be determined using the cluster chain obtained by breaking the cycle at S.

For example, whether belief updating is achievable by message passing in the cluster graphs in Figure 3.15(a) and (c) can be determined by deleting the separator $\{x\}$ in (a) to obtain (b), and deleting any separator $\{x\}$ in (c) to obtain, say, (d). Belief updating by message passing is achievable in (b), and we will show how in Section 5.5. On the other hand, belief updating by message passing is *not* achievable in (d), and we will explain why in Section 3.7. The key conclusion is that the cyclic structure of a weak degenerate cycle is insignificant just as that of a strong degenerate cycle is. Hence, a cluster graph with only degenerate cycles can always be treated by first breaking the cycles at appropriate separators. The resultant is a cluster tree.

Given a connected graph G, its connected subgraphs with the same nodes as G and the minimum number of links are trees. That is, trees are the simplest subgraphs

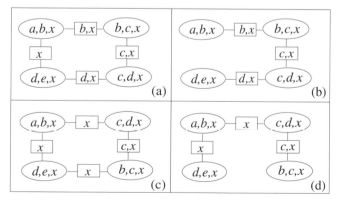

Figure 3.15: Weak degenerate cycles in (a) and (c) are broken into chains in (b) and (d).

that retain connectedness. Simplicity is conducive to efficiency. For message passing, a cluster tree is more efficient to maintain than the corresponding cluster graph
with degenerate cycles. There are less separators to maintain at the least.

Because a cluster graph with nondegenerate cycles cannot support correct belief
updating, whereas a cluster graph with degenerate cycles can always be substituted
with a simpler cluster tree, we will focus on using cluster tree structures to support
belief updating through message passing.

3.7 Junction Trees

We have identified cluster trees as the general cluster graph structure for belief
updating by message passing. We now consider if additional restrictions to the tree
structure are necessary. In Section 3.5, we introduced the notion of consistency
between a pair of adjacent clusters and that of a locally consistent cluster graph.
First we extend the notion of consistency to a larger scale.

Consider a cluster tree H over a domain V. Pick a variable $v \in V$ that is contained
in two or more clusters. From each such cluster Q with a potential $B_Q(Q)$, the
distribution $P_Q(v)$ of v can be obtained as

$$P_Q(v) = \text{const} \sum_{Q \setminus \{v\}} B_Q(Q).$$

Suppose $v = v_0$ is observed. After belief updating, we would expect $P(v = v_0) =
1.0$ in each cluster that contains v. If v is not observed, we would expect the prior
distribution $P(v)$ to be identical in those clusters. If other variables are observed,
we would expect the posterior distribution $P(v|\text{obs})$ to be identical after belief
updating, where obs denotes the observation.

We expect the preceding results because $P(v)$ represents the belief of a single
agent no matter which cluster it is associated with, and a rational agent should not
contradict itself. Such a state of belief is formally defined as follows: If each pair
of clusters Q_1 and Q_2 ($Q_1 \cap Q_2 \neq \emptyset$) in a cluster tree H satisfies

$$\sum_{Q_1 \setminus Q_2} B_{Q_1}(Q_1) = \text{const} \sum_{Q_2 \setminus Q_1} B_{Q_2}(Q_2),$$

H is *globally consistent*. Note that global consistency is applicable to both prior
and posterior potentials.

Does local consistency ensure global consistency? The answer is no, as illustrated
in Figure 3.16. Suppose each variable is binary with space $\{0, 1\}$. The potential
of each cluster has the value 1.0 for one configuration as shown and 0 for the
other configurations. The cluster tree is locally consistent because Q_1 and Q_2 have
identical $P(b)$ and Q_2 and Q_3 have identical $P(c)$. However, it is *not* globally

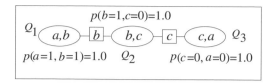

Figure 3.16: A locally consistent but globally inconsistent cluster tree.

consistent, for $P_{Q_1}(a = 1) = 1.0$ but $P_{Q_3}(a = 1) = 0$. The same problem can also occur in the cluster tree of Figure 3.15(d).

Local consistency can be achieved effectively by message passing in a cluster tree. On the other hand, global consistency is required by the semantics of a cluster tree. The preceding example illustrates that message passing along the separators of an arbitrary cluster tree does not ensure global consistency. In order to achieve global consistency through local consistency, the cluster trees must be *junction trees*, as we will show in Definition 3.6.

Definition 3.6 *A cluster tree is a **junction tree** (JT) if for every pair of clusters Q_1 and Q_2 and the path ρ between them, $Q_1 \cap Q_2$ is contained in each cluster on ρ.*

Two representative junction trees are shown in Figure 3.17. In (a), the intersection between nonadjacent clusters Q_1 and Q_3 is empty and hence the cluster tree is a JT. In (b), the intersection between Q_2 and Q_4, namely $\{e\}$, is contained in Q_3, and so is the intersection between Q_1 and Q_4.

The importance of JT structures is established in the following theorem:

Theorem 3.7 *Let H be a cluster tree representation of a domain V.*

1. *If H is a junction tree, H is globally consistent whenever it is locally consistent.*
2. *If H is not a junction tree, then there exist potentials for clusters such that H is locally consistent but not globally consistent.*

Proof:

(1) We prove by induction on the length L of the longest path in H. Let H be a locally consistent JT. If $L = 1$, H is globally consistent. Assume that H is globally consistent when $L = k$.

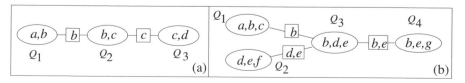

Figure 3.17: Two junction trees.

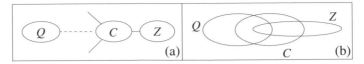

Figure 3.18: Illustration of proof for Theorem 3.7 (1).

Consider $L = k + 1$. Let the two end clusters of a longest path in H be Q and Z, and let C be the cluster adjacent to Z on the path, as shown in Figure 3.18(a). The relation among the three sets Q, C, and Z is shown in the Venn Diagram in (b). Note that $Q \cap Z$ is contained in C because H is a JT.

The length from Q to C is k. By assumption, Q and C are consistent, and we have

$$B_Q(Q \cap Z) = \sum_{Q \cap C \setminus Z} B_Q(Q \cap C) = \text{const} \sum_{Q \cap C \setminus Z} B_C(Q \cap C)$$

$$= \text{const} \sum_{Q \cap C \setminus Z} \sum_{C \setminus Q} B_C(C) = \text{const} \sum_{C \setminus (Q \cap Z)} B_C(C).$$

Because H is locally consistent, C and Z are consistent, and we have

$$B_Z(Q \cap Z) = \sum_{C \cap Z \setminus Q} B_Z(C \cap Z) = \text{const} \sum_{C \cap Z \setminus Q} B_C(C \cap Z)$$

$$= \text{const} \sum_{C \cap Z \setminus Q} \sum_{C \setminus Z} B_C(C) = \text{const} \sum_{C \setminus (Q \cap Z)} B_C(C).$$

Hence, we have

$$B_Q(Q \cap Z) = \text{const } B_Z(Q \cap Z).$$

(2) Suppose H is not a JT. Then there exist two clusters Q and Z and a third cluster W on the path between them such that $Q \cap Z$ is not contained in W. That is, we have $X = (Q \cap Z) \setminus W \neq \emptyset$. Let C be the cluster adjacent to W on the path between Q and W, as shown in Figure 3.19. Note that it is possible that $C = Q$. Denote the subtree rooted at C by T_Q and the subtree rooted at W by T_Z, as shown by the dashed dividing line in Figure 3.19.

We now construct a potential for each cluster in H. A variable may occur in more than one cluster. First, we assign a value for each occurrence of each variable. For each $v \in V \setminus X$, assign $v = v_0$ to each of its occurrences. For each $x \in X$, assign

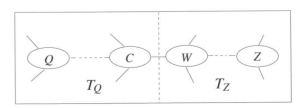

Figure 3.19: Illustration of proof for Theorem 3.7 (2).

$x = x_0$ to each of its occurrences in T_Q, but assign $x = x_1$ to each of its occurrences in T_Z. Note that each x has at least one occurrence in T_Q (e.g., Q) and one occurrence in T_Z (e.g., Z), but it does not occur in cluster W.

Next, we set the potential in each cluster. For each cluster Y and each configuration y, if there exists $u \in Y$ such that its value in y differs from the value assigned to the occurrence of u in Y, then set $B_Y(y) = 0$. Otherwise, set $B_Y(y) = 1$. Note that only a single configuration in each cluster has the nonzero potential value.

We claim that with potentials thus determined, H is locally consistent: T_Q is locally consistent because, for each pair of adjacent clusters, their configurations with nonzero potential values are compatible. The same is true for T_Z. Clusters W and C are consistent because $W \cap X = \emptyset$; hence, $W \cap C \subset (V \backslash X)$. Therefore, their configurations with nonzero potential values are compatible.

Clearly, H is not globally consistent because, for each $x \in X$, $P_Q(x_0) = 1$ and $P_Z(x_0) = 0$. $\qquad\qquad\qquad\qquad\qquad\qquad\qquad\qquad\qquad\qquad\qquad\qquad\qquad\qquad$ $\square$

Theorem 3.7 formally states that, in order to perform belief updating by message passing, the domain variables must be organized into a junction tree. Given that a JT structure must be used in general in order to achieve global consistency through local consistency, we investigate in Chapter 4 how to convert a BN into a suitable JT representation for inference.

3.8 Bibliographical Notes

A preliminary study of message passing in degenerate and nondegenerate cycles is found in Xiang and Lesser [92] in the context of cooperative multiagent systems. The analysis presented there has been extended significantly in this chapter for justifying cluster-tree-based message passing. Theorem 3.7 (1) is taken from Jensen, Lauritzen, and Olesen [30]. A special case of Theorem 3.7 (2), which has been generalized here, was presented in Jensen [29].

Using message passing in junction tress for belief updating has been studied by many, including Spiegelhalter [67]; Lauritzen and Spiegelhalter [37]; Shafer et al. [64]; Jensen et al. [30]; and Madsen and Jensen [39]. The use of junction trees in database management can be found in Maier [40], where they are referred to as *join trees*.

3.9 Exercises

1. Describe the cluster graph in Figure 3.5 using the (V, Ω, E) notation.
2. Compare the computation performed in Table 3.1 and that in Tables 2.2, 2.3, and 2.4. If the intermediate results in Table 3.1 were represented as probability distributions, what would be the normalizing constants?

3. For the example in Section 3.5, verify that messages $P(b|d_0)$ and $P(c|d_0)$ on the non-degenerate cycle computed from *jpd* are identical to those computed from *jpd'*.

4. For the example in Section 3.5, verify that $P(a|d_0)$ computed from *jpd* by correct belief updating should be different from that computed from *jpd'*.

5. Generalize Lemmas 3.2 and 3.3 and Theorem 3.4 on strong nondegenerate cycles to the case in which each of a, b, c, d is a set of variables.

6. Demonstrate that belief updating by message passing is not possible in general in the cluster graph of Figure 3.15(d).

7. Let the domain of a BN be $V = \{a, b, c, d\}$. Find all cyclic junction graphs of three clusters such that none of the clusters is a subset of another. Determine the ratio between those junction graphs with degenerate cycles and those with nondegenerate cycles.

8. Determine if the cluster graphs in Figure 3.15 are junction trees.

9. From the generating set $V = \{a, b, c, d, e, f, g, h, i, j\}$ construct a cluster tree that has five clusters and is not a chain.

10. Check if the cluster tree created in Exercise 9 is a JT. If not, modify the membership of clusters to make it a JT.

11. Assign potentials to clusters in the JT created in Exercise 9 such that the JT is locally consistent. Verify its global consistence.

12. Prove that after several adjacent clusters of a JT are merged, the resultant cluster graph is still a JT.

4

Junction Tree Representation

Chapter 3 has shown that, in order to use concise message passing in a single cluster graph for exact belief updating with a nontree BN, one must reorganize the DAG into a junction tree. Graphical representations of probabilistic knowledge result in efficiency through the exploration of conditional independence in terms of graphical separation, as seen in Chapter 2. Therefore, the reorganization needs to preserve the independence–separation relations of the BN as much as possible. This chapter formally describes how independence is mapped into separation in different graphical structures and presents algorithms for converting a DAG dependence structure into a junction tree while preserving graphical separation to the extent possible.

Section 4.2 defines the graphical separation in three types of graphs commonly used for modeling probabilistic knowledge: *u-separation* in undirected graphs, *d-separation* in directed acyclic graphs, and *h-separation* in junction trees. The relation between conditional independence and the sufficient content of a message in concise message passing is established in Section 4.3. In Section 4.4, the concept of the *independence map* or *I-map*, which ties a graphical model to a problem domain based on the extent to which the model captures the conditional independence of the domain, is introduced. The concept of a *moral graph* is also introduced as an intermediate undirected graphical model to facilitate the conversion of a DAG model to a junction tree model. Section 4.5 introduces a class of undirected graphs known as *chordal graphs* and establishes the relation between chordal graphs and junction trees. It is shown that a moral graph model must be converted to a chordal graph in order to construct a junction tree model. The expressiveness of chordal graphs and junction trees in representing conditional independence is shown to be equivalent in Section 4.7. Section 4.6 presents an algorithm known as *elimination* for converting a moral graph into a chordal graph, and Section 4.8 describes algorithms to convert a chordal graph into a junction tree.

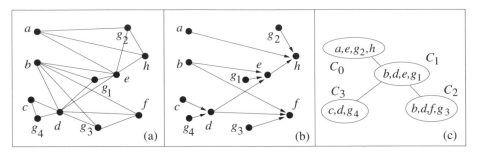

Figure 4.1: (a) An undirected graph. (b) A DAG. (c) A junction tree.

4.1 Guide to Chapter 4

The essential benefit from graphical representation of probabilistic knowledge is the ability to specify only what is directly relevant. Graphs provide this ability through graphical connection and separation. What is directly connected is directly relevant. What is separated is irrelevant given the separating variables. Although graphical connection is explicitly indicated by links in the graph, graphical separation is less obvious. Section 4.2 specifies the criteria of graphical separation for three types of graphs. For undirected graphs such as Figure 4.1(a), separation is determined by path blocking. For example, the node a is said to be separated from b by e because e blocks the path between a and b. Separation can occur between groups of nodes as well. For instance, nodes e, g_1 are separated from nodes f, g_3 by nodes b, d. The separation criterion for undirected graphs is termed *u-separation*.

For directed acyclic graphs such as Figure 4.1(b), separation is determined by path analysis that takes into account the directions of arcs. In Figure 4.1(b), a is separated from b by e. However, a is *not* separated from b by h. The reason is that the path between a and b has arcs pointing at h. The rationale is that h represents an effect or consequence of the causes a and b. Once the effect is known, the two causes compete to explain the effect and hence become relevant. The separation criterion for directed acyclic graphs is termed *d-separation*.

For junction trees such as that of Figure 4.1(c), separation is defined between elements of clusters. The elements a and b are separated by e because e is the separator between the cluster C_0 (containing a) and the clusters C_1 and C_2 (containing b). The separation criterion for junction trees is termed *h-separation*.

Having defined criteria for graphical separation, we move to the issue of relevance and irrelevance among variables (represented by nodes in the graph) in Section 4.3, which is concerned with the following question: When a message is to be passed between two groups of variables, what is the necessary and sufficient message to update the belief at the destination group? A necessary and sufficient message would make additional information about the source group irrelevant. The irrelevance can be described using the terminology of conditional independence. The conclusion drawn is the following: If the variables shared by the two groups render them conditionally independent, then the belief in the shared variables is the sufficient message. Otherwise, the belief in the shared variables is an insufficient message.

Section 4.4 ties graphical separation and conditional independence together. An ideal graph for encoding the knowledge about a problem domain is one in which graphical separation always implies conditional independence. That is, the graph is so connected that whenever two groups of nodes are separated by a third group, the two corresponding groups of variables become irrelevant once the value of the third group of variables is known. Hence, passing the belief of the third group between the two groups is sufficient for belief updating. Such a graph for a given problem domain is called an *I-map*, which allows effective representation of probabilistic knowledge and effective belief updating by concise message passing.

Chapter 3 has established that belief updating in nontree BNs can be achieved by concise message passing only in junction tree representations. As the first step in converting a BN into a junction tree, Section 4.4 also presents the procedure for converting a DAG I-map (the structure of a BN) into an undirected I-map called a *moral* graph. The conversion essentially connects parents of each node and then drops the directions of arcs. For example, if the DAG of a BN is that of Figure 4.1(b), then its moral graph is that of Figure 4.1(a).

Section 4.5 investigates the next step of conversion. It raises the issue that, for some moral graphs, no junction trees can be found. Therefore, to ensure the conversion of a given moral graph into a junction tree, additional processing called *triangulation* has to be performed on the moral graph. The process essentially adds some links to the moral graph, and the resultant is called a *chordal* graph. Figure 4.1(a) is a "lucky" moral graph that is already chordal. Figure 4.2(b) shows an "unlucky" moral graph obtained from the DAG in (a). After a link is added, as shown in (c), the moral graph becomes chordal. Intuitively, the square in (b) has been turned into triangles, and hence the name *triangulation*.

Section 4.6 explains a simple method called *node elimination* to triangulate any moral graphs. The method removes nodes in the moral graph one by one. Before removal of each node, it makes sure that nodes in the neighborhood are pairwise connected. For example, if the nodes in Figure 4.2(b) are removed in the order

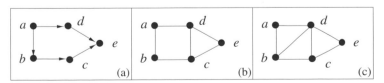

Figure 4.2: (a) A DAG. (b) The moral graph of (a). A chordal graph obtained from (b).

(a, b, d, c, e), then a link $\langle b, d \rangle$ needs to be added, which gives rise to the chordal graph in (c).

Adding links during the conversion from DAGs to moral graphs and then to chordal graphs exacts a price. The conversion destroys some graphical separations and makes some conditional independence invisible. Before going into the last step of the conversion, from the chordal graph into a junction tree, Section 4.7 tells us that this last step is free from such expense.

Section 4.8 deals with the details of the last conversion step. The clusters of the junction tree are first defined. Each cluster comes from a group of pairwise-connected nodes in the chordal graph. For instance, the junction tree in Figure 4.1(c) is converted from the chordal graph in (a). The elements of cluster Q_0 in (c) are pairwise connected in (a). Afterwards, each pair of clusters sharing some elements is connected to form a cluster graph. Finally, some connections are deleted to turn the cluster graph into a tree. The graphical structure of the junction tree representation is then completed.

4.2 Graphical Separation

The fundamental property of graphs for effective probabilistic inference is graphical separation. Graph separation is based on a path analysis. Let X, Y, and Z be disjoint subsets of nodes in a graph G (directed or undirected). A path between nodes $x \in X$ and $y \in Y$ is rendered *closed* (or *blocked*), or *open* (or *active*) by Z. If every path between every pair of x and y is closed, then X and Y are said to be *separated*. On the other hand, if there exists one open path between a pair x and y, then X and Y are not separated. What condition renders a path closed or open differs between directed and undirected graphs. These concepts are presented precisely in the paragraphs that follow.

Let $G = (V, E)$ be an undirected graph. A path ρ between $x \in X$ and $y \in Y$ is closed by Z if there exists $z \in Z$ on ρ. Otherwise, ρ is rendered open by Z.

Intuitively, each node on a path is like a valve in a pipe. The default state of the valve is open. A set of valves whose default states are reversed (from open to closed) is denoted by Z. If one valve on a pipe is closed, then the pipe is closed.

Two nodes x and y are *separated* by a set Z of nodes if every path between x and y is closed by Z; X and Y are said to be *separated* by Z if for every $x \in X$ and $y \in Y$, x and y are separated by Z; $\langle X|Z|Y \rangle_G$ is used to denote that X and Y are separated by Z in G. When there is no confusion, we write $\langle X|Z|Y \rangle$. When a set is a singleton, say $|X| = 1$, we write $\langle x|Z|Y \rangle$ instead of $\langle \{x\}|Z|Y \rangle$ for simplicity. In Figure 2.4(a), we have $\langle v_5|v_2|v_3 \rangle$ and $\langle v_5|\{v_2, v_1\}|v_3 \rangle$, but $\neg \langle v_5|v_1|v_3 \rangle$.

When G is a DAG, $\langle X|Z|Y \rangle_G$ takes into account the direction of arcs. The corresponding criterion is called *d-separation*. The criterion defined above will be

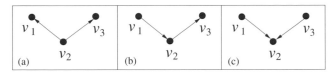

Figure 4.3: The node v_2 is tail-to-tail in (a), head-to-tail in (b), and head-to-head in (c).

referred to as *u-separation*, where *u* stands for *undirected*. Before we proceed to d-separation, the u-separation is summerized as follows:

Definition 4.1 *Let G be an undirected graph and X, Y, Z be disjoint sets of nodes in G. A path ρ between nodes $x \in X$ and $y \in Y$ is **closed** by Z if there exists $z \in Z$ on ρ. Otherwise, ρ is rendered **open** by Z.*

*Nodes x and y are **u-separated** by Z if every path between x and y is closed by Z.*

*X and Y are **u-separated** by Z if for every $x \in X$ and $y \in Y$, x and y are u-separated by Z.*

In a directed acyclic graph, when two arcs meet in a path, the node shared by the arcs can be described as: *tail-to-tail, head-to-tail,* or *head-to-head*, as shown for the node v_2 in Figure 4.3. The d-separation criterion for $\langle X|Z|Y \rangle$ is identical to that for u-separation if each path between an x and a y contains no head-to-head nodes. A slight complication arises when some paths have head-to-head nodes. The d-separation is defined precisely in the paragraphs that follow.

A path ρ between nodes x and y is closed by Z whenever one of the following two conditions is true: (1) There exists $z \in Z$ that is either tail-to-tail or head-to-tail on ρ. (2) There exists a node v that is head-to-head on ρ, and neither v nor any descendant of v is in Z. If both conditions are false, then ρ is rendered open by Z.

From the standpoint of the pipe and valve analogy, each tail-to-tail or head-to-tail node on a path is like a valve whose default state is open. Each head-to-head node is like a valve whose default state is closed. A set of valves whose default states are reversed is denoted by Z. For a head-to-head node, if either it or a descendant is in Z, the state of the valve is reversed.

Two nodes x and y are d-separated by Z if every path between x and y is closed by Z; X and Y are d-separated by Z if for every $x \in X$ and $y \in Y$, x and y are d-separated by Z. In Figure 4.4, $\langle a|\{f, d\}|h \rangle$ holds because f is head-to-tail in both paths from a to h. It is also true $\langle b|\emptyset|d \rangle$ because the upper path from b to d has a closed valve e and the lower path has a closed valve f. The statement $\langle c|d|e \rangle$ is false because, although the path from c to e through f is closed, the other path is open. The d-separation is summarized as follows:

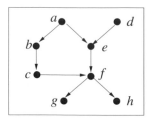

Figure 4.4: Illustration of d-separation.

Definition 4.2 *Let G be a directed acyclic graph and X, Y, Z be disjoint sets of nodes in G. A path ρ between nodes $x \in X$ and $y \in Y$ is **closed** by Z if one of the following two conditions holds: (1) There exists $z \in Z$ that is either tail-to-tail or head-to-tail on ρ. (2) There exists a node v that is head-to-head on ρ and neither v nor any descendant of v is in Z. If both conditions fail, then ρ is rendered **open** by Z.*

*Nodes x and y are **d-separated** by Z if every path between x and y is closed by Z;*

*X and Y are **d-separated** by Z if for every $x \in X$ and $y \in Y$, x and y are d-separated by Z.*

Because the JT representation of BNs is used for inference, we define $\langle X|Z|Y \rangle_H$ in a JT H over V, where $X, Y, Z \subset V$; $\langle X|Z|Y \rangle_H$ will be referred to as *h-separation*. Let $x \in X$ and $y \in Y$ be contained in distinct clusters Q_x and Q_y, respectively, and no cluster in H contains both x and y. Because H is a tree, the path ρ between Q_x and Q_y is unique. It is closed by Z if there exists a separator $S \subseteq Z$ on ρ. Otherwise, ρ is rendered open by Z. The following proposition shows that if ρ is closed by Z, then every path between a cluster containing x and a cluster containing y is closed.

Proposition 4.3 *Let H be a JT over V. Let $\{x\}$, $\{y\}$, Z be disjoint subsets of V so that x and y are contained in distinct clusters Q_x and Q_y, and no cluster contains both x and y. If the path ρ between Q_x and Q_y is closed by Z, then every path between a cluster containing x and a cluster containing y is closed.*

Proof: We prove by contradiction. Suppose that there exist two clusters Q'_x and Q'_y such that (1) $x \in Q'_x$ and $y \in Q'_y$ hold, (2) it is not the case that both $Q'_x = Q_x$ and $Q'_y = Q_y$, and (3) the path ρ' between Q'_x and Q'_y is rendered open by Z. By definition, (3) means that there exists no separator S' exists on ρ' such that $S' \subseteq Z$. Hence, ρ cannot be part of ρ'. This leads to the possible relations between ρ and ρ', as shown in Figure 4.5.

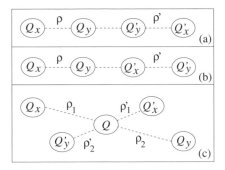

Figure 4.5: Relation between paths ρ and ρ' in proof of Proposition 4.3.

The paths ρ and ρ' may be noncrossing as in (a), (b), and two other symmetric cases (not shown), or they may cross each other at a cluster Q as in (c), where $\rho_1 + \rho_2 = \rho$ and $\rho_1' + \rho_2' = \rho'$. Each dashed line signifies possible additional clusters on the path. Note that it is possible $Q_y = Q_y'$ in (a). Otherwise, Q_y and Q_y' are connected in (a) outside ρ and ρ' because H is a tree, and Q_y and Q_x' are similarly connected in (b). Consider case (a). Denote the path between Q_x and Q_x' by ω. Because ρ is closed by Z and ρ is part of ω, there exists a separator $S \subseteq Z$ in ω. Because $x \notin Z$, a separator S has been found between Q_x and Q_x' such that $x \notin S$. This implies that H is not a JT, which is a contradiction. The same can be shown similarly for case (b).

Next, consider case (c) in Figure 4.5. Because ρ is closed by Z, there exists a separator $S \subseteq Z$ in ρ. Because $\rho_1 + \rho_2 = \rho$, S is either in ρ_1 or in ρ_2. If S is in ρ_1, define $\omega = \rho_1 + \rho_1'$. A separator S has been found on ω between Q_x and Q_x' such that $x \notin S$. If S is in ρ_2, define $\omega = \rho_2 + \rho_2'$. A separator S has been found on ω between Q_y and Q_y' such that $y \notin S$. Either way, it implies that H is not a JT, which is a contradiction. $\qquad\square$

On the basis of Proposition 4.3, when the path ρ between Q_x and Q_y is closed by Z, it is said that x is *h-separated* from y by Z, or $\langle x|Z|y \rangle$, and X and Y are *h-separated* by Z if $\langle x|Z|y \rangle$ for every pair of x and y. In Figure 3.17(b), we have $\langle a|b|d \rangle$, $\neg\langle a|\emptyset|d \rangle$, and $\langle \{a, f\}|\{b, e\}|g \rangle$. The h-separation is summarized as follows:

Definition 4.4 *Let H be a JT over V, and X, Y, Z be disjoint subsets of V such that no $x \in X$ and $y \in Y$ are contained in the same cluster in H. For $x \in X$ contained in cluster Q_x and $y \in Y$ contained in cluster Q_y, x and y are **h-separated** by Z if there exists a separator $S \subseteq Z$ exists on the path between Q_x and Q_y. Otherwise, x and y are not h-separated by Z.*

*Moreover, X and Y are **h-separated** by Z if for every $x \in X$ and $y \in Y$, x and y are h-separated by Z.*

Note that if there exists $x \in X$ and $y \in Y$ that are contained in the same cluster in H, then X and Y are not h-separated by Z. In fact, they cannot be h-separated at all. The h-separation could have been defined over general cluster graphs instead of over junction trees, as in Definition 4.4. Such generality, however, is not needed for the purposes of this book.

4.3 Sufficient Message and Independence

To perform effective probabilistic inference in junction trees, we pass concise messages over separators between adjacent clusters. Under what condition are the messages sufficiently informative to ensure correct inference? Consider two adjacent clusters $C = X \cup Z$ and $Q = Y \cup Z$, where X, Y, and Z are disjoint and Z is the separator. Suppose each cluster is associated with a potential $P_C(X, Z) = \sum_Y P(X, Y, Z)$ and $P_Q(Y, Z) = \sum_X P(X, Y, Z)$. If $I(X, Z, Y)$ holds, then $P(X, Y, Z) = P_C(X, Z)P_Q(Y, Z)/P_C(Z)$, where $P_C(Z) = \sum_X P_C(X, Z) = \sum_Y P_Q(Y, Z) = P_Q(Z)$.

Suppose some variables in C are observed. Using the method in Section 2.3, update belief in C to get $P_C(X, Z|\text{obs})$. To update belief in Q, pass the message $P_C(Z|\text{obs}) - \sum_X P_C(X, Z|\text{obs})$ from C to Q and replace the belief in Q by

$$P_Q(Y|Z) * P_C(Z|\text{obs}) = P_Q(Y, Z|\text{obs}).$$

The message passing is correct because

$$P(X, Y, Z|\text{obs}) = \frac{P_C(X, Z|\text{obs})P_Q(Y, Z|\text{obs})}{P_C(Z|\text{obs})}.$$

What if $I(X, Z, Y)$ does not hold? Consider the graph G in Figure 3.12(a). If a BN is defined with the dependence structure G and $P(a, b, c, d)$ is constructed by chain rule, then in general, $I(a, b, \{c, d\})$ does not hold (see Exercise 3). It has been shown in Lemma 3.2 that passing a message over b from cluster $\{b, c, d\}$ to $\{a, b\}$ cannot produce a correct posterior in general. This is summarized in the following proposition:

Proposition 4.5 *Let X, Y, and Z be disjoint sets of variables with $P(X, Y, Z)$ defined. Let $C = X \cup Z$ be associated with $P_C(X, Z) = \sum_Y P(X, Y, Z)$ and $Q = Y \cup Z$ be associated with $P_Q(Y, Z) = \sum_X P(X, Y, Z)$.*

1. *If $I(X, Z, Y)$ holds, then belief updating can be performed correctly by passing a potential over Z between C and Q.*

2. *If $I(X, Z, Y)$ does not hold, then belief updating cannot be performed correctly in general by passing a potential over Z between C and Q.*

To conclude, in general, when $I(X, Z, Y)$ holds, the sufficient amount of message to be passed between C and Q is the potential over Z. Hence, to organize message passing effectively, the graphical structure should encode the conditional independence explicitly. To use a JT representation of a BN for inference, the JT must preserve the conditional independence as much as possible in the original BN.

4.4 Encoding Independence in Graphs

The idea of encoding conditional independence in graphs is formally described by the following notion of independence map [52]:

Definition 4.6 *A graph (directed or undirected) G is an **independence map** or **I-map** of a domain V if there is a one-to-one correspondence between nodes of G and variables in V and $\langle X|Z|Y \rangle_G$ implies $I(X, Z, Y)$ for all disjoint subsets X, Y, and Z of V.*

By definition, adding additional links to an I-map does not change its I-mapness. Such links make some conditional independence graphically invisible and hence should be avoided. A *minimal I-map* $G = (V, E)$ is an I-map such that no subgraph $G' = (V, E')$ $(E' \subset E)$ is also an I-map. Equipped with these concepts, we can restate what is required in a graphical structure: To organize concise message passing effectively, a graphical structure should be a minimal I-map.

Consider the DAG structure G in Figure 2.6(b), which is a minimal I-map with respect to our knowledge about the circuit in (a). The arc (g_2, h) represents the direct causal dependence of output h on the state of gate g_2. The absence of an arc from c to f and the d-separation $\langle c|d|f \rangle$ signify that, once we know the value of d, our belief on the value of f is unaffected by our knowledge of the value of c, or $I(c, d, f)$. Note that d is a head-to-tail node on the path from c to f. Similarly, the d-separation $\langle e|\{b, d\}|f \rangle$ expresses that, once we know the value of b and d, knowing in addition the value of e does not change our belief in the value of f, or $I(e, \{b, d\}, f)$. Note that b and d are tail-to-tail nodes on their paths from e to f. The absence of an arc between c and g_4 and the d-separation $\langle c|\emptyset|g_4 \rangle$ signify that normally the value of input c and the state of gate g_4 are irrelevant to each other, or $I(c, \emptyset, g_4)$. However, $\neg \langle c|d|g_4 \rangle$ does not allow us to infer $I(c, d, g_4)$. In fact, $I(c, d, g_4)$ is false because once we know the output d, the knowledge about c will allow us to infer the state of g_4. Note that d is a head-to-head node on the path

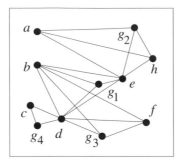

Figure 4.6: Moral graph for the DAG in Figure 2.6(b).

from c to g_4. The dependence between c and g_4 after knowing d is called *induced dependence* (Pearl [52]).

Alternatively, the knowledge about the same circuit may be represented through an undirected graph. Examine the relationship between the preceding DAG I-map G and an undirected I-map G' of the same circuit. One candidate for G' is a graph G^* obtained from the DAG G in Figure 2.6(b) by dropping the direction of each arc. Graph G^* is called the *skeleton* of G. However, G^* is not an I-map. For example, $\langle c|d|g_4 \rangle$ holds in G^*, whereas $I(c, d, g_4)$ is known to be false. This is because, without directions, an undirected graph is unable to represent $\langle c|\emptyset|g_4 \rangle$ and $\neg \langle c|d|g_4 \rangle$ simultaneously. To negate $\langle c|d|g_4 \rangle$, connect c and g_4 with a link in G'. For the same reason, for each child node of G, add a link in G' between each pair of its parents. The resultant G' is shown in Figure 4.6. In general, the graph resulting from such processing is called a *moral graph* as defined below:

Definition 4.7 *Let G be a DAG. For each child node in G, connect its parent nodes pairwise and drop the direction of each arc. The resultant undirected graph G' is the **moral graph** of G.*

The graph in Figure 4.6 is the moral graph for the DAG in Figure 2.6(b). In Theorem 4.8 it is shown that, given a DAG as a minimal I-map, a minimal undirected I-map is its moral graph.

Theorem 4.8 *Let a DAG G be a minimal I-map over V and G' be its moral graph. Then G' is a minimal I-map over V.*

Proof: Let G^* be the skeleton of G. Because G is a minimal I-map and a link $\langle x, y \rangle$ renders x and y inseparable under both d-separation and u-separation, an undirected I-map must contain at least links of G^*.

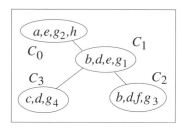

Figure 4.7: JT structure for the DAG in Figure 2.6(b).

The only reason that G^* is not an I-map is because, for every pair of nodes x and y that has a common child c in G, if $\langle x, y \rangle$ is not in G (hence not in G^*), then there exists a set of nodes Z that includes c such that $\langle x|Z|y \rangle_{G^*}$ whereas $\neg I(x, Z, y)$. Because $\langle x|Z|y \rangle$ no longer holds in G', G' is an I-map. Because each link $\langle x, y \rangle$ added to G^* negates $\langle x|Z|y \rangle$ for all suitable instances of Z and no addition of other links does so, the additional links are minimal. $\qquad\square$

Because the JT representation will be used for message passing, we extend the concept of I-map to JTs in Definition 4.9.

Definition 4.9 *A junction tree H over V is an I-map if $\cup_i Q_i = V$, where each Q_i is a cluster in H, and $\langle X|Z|Y \rangle_H$ implies $I(X, Z, Y)$ for all disjoint subsets X, Y, and Z of V.*

As an example, Figure 4.7 shows a JT for the circuit in Figure 2.6.

Assuming that the original (versus converted or compiled) knowledge representation is in the form of a BN, we need to compile it into a JT representation. The compilation, as discussed earlier in this section as well as in the preceding section, must preserve the I-mapness. How to perform such a compilation is the topic of the next section.

4.5 Junction Trees and Chordal Graphs

A JT I-map can be derived from the moral graph of the DAG rather than from the DAG directly. This is because a JT uniquely defines an undirected graphical structure, and hence a moral graph (as a minimal undirected I-map) of the DAG provides a more direct basis to work on than the DAG I-map itself. For example, consider the JT in Figure 4.7. Construct an undirected graph G with the generating set of the JT as the nodes. For each pair of nodes contained in a cluster in the JT, connect the pair in G. The resultant G is the graph in Figure 4.6.

To construct a JT, the clusters must be determined. Which components of a moral graph correspond to a cluster? In a JT, each cluster admits no graphical separation

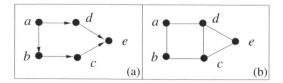

Figure 4.8: (a) A DAG. (b) The moral graph of (a).

(Definition 4.4) and signifies no conditional independence internally. Similarly, in an undirected graph, a set of pairwise-connected nodes admits no graphical separation (Definition 4.1).

A set of nodes in an undirected graph G is *complete* if they are pairwise connected. A maximal set of nodes that is complete is called a *clique* (or *maximal clique*) in G. In other words, if C is a clique in G, then no proper superset of nodes in G is also a clique. There are exactly four cliques in Figure 4.6:

$$C_0 = \{a, e, g_2, h\}, \quad C_1 = \{b, d, e, g_1\}, \quad C_2 = \{b, d, f, g_3\}, \quad C_3 = \{c, d, g_4\}.$$

Because a clique is a maximal set of variables without graphically identifiable conditional independence, it should become a cluster in the JT representation. The JT in Figure 4.7 has C_0 through C_3 as its clusters.

Can we construct a JT from cliques of every moral graph? Unfortunately we cannot. Consider the DAG in Figure 4.8(a) with its moral graph in (b). The cliques are

$$C_0 = \{a, b\}, \quad C_1 = \{a, d\}, \quad C_2 = \{b, c\}, \quad C_3 = \{c, d, e\}.$$

No cluster graph made out of these clusters is a JT (Exercise 4). In Theorem 4.10 we present the condition that ensures the existence of a JT made out of the cliques of an undirected graph.

Consider an undirected graph G. A path or cycle ρ has a *chord* if there is a link in G between two nonadjacent nodes on ρ. G is *chordal* or *triangulated* if every cycle of length greater than or equal to 4 has a chord. A cycle of length ≥ 4 without a chord is a *chordless* cycle. Figure 4.6 is chordal, but Figure 4.8(b) is not because the cycle $\langle a, b, c, d, a \rangle$ of length 4 does not have a chord.

Theorem 4.10 *Let $G = (V, E)$ be a connected undirected graph and Ω be the set of cliques of G. Then there exists a junction tree whose clusters are elements of Ω if and only if G is chordal.*

To prove the theorem, the following lemmas are needed. We first introduce necessary concepts for the lemmas.

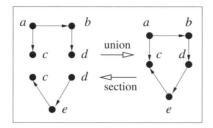

Figure 4.9: Two connected graphs on the left are "unioned" into the graph on the right.

Definition 4.11 *Let $G_i = (V_i, E_i)$ $(i = 0, 1)$ be two graphs (directed or undirected), G_0 and G_1 are said to be **graph consistent** if the subgraphs of G_0 and G_1 spanned by $V_0 \cap V_1$ are identical.*

*Given two graphs $G_i = (V_i, E_i)$ $(i = 0, 1)$ that are consistent, the graph $G = (V_0 \cup V_1, E_0 \cup E_1)$ is called the **union** of G_0 and G_1 denoted by $G = G_0 \sqcup G_1$.*

Although graph union does not require graph consistency in general, this book only uses the operation for consistent graphs.

Definition 4.12 *Given a graph $G = (V, E)$, V_0 and V_1 such that $V_0 \cup V_1 = V$, and subgraphs G_i of G spanned by V_i $(i = 0, 1)$ such that $G = G_0 \sqcup G_1$, G is said to be **sectioned** into G_0 and G_1.*

Although the definition allows the case $V_0 \cap V_1 = \emptyset$, this book concerns mainly the case $V_0 \cap V_1 \neq \emptyset$. Note that if G_0 and G_1 are sectioned from a third graph, then G_0 and G_1 are graph consistent. Figure 4.9 gives one example of directed graphs. The union of multiple graphs or the sectioning of a graph into multiple graphs can be similarly defined. An example is given in Figure 4.10 for undirected graphs.

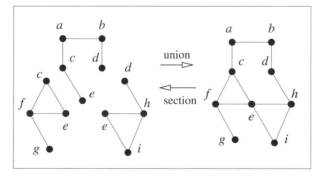

Figure 4.10: The graph on the right is sectioned into three connected graphs on the left.

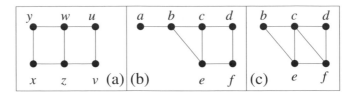

Figure 4.11: Illustration of simplicial nodes.

Lemma 4.13 *Let $G = (V, E)$ ($|V| > 3$) be a noncomplete, connected chordal graph. Then there exists a complete subset $Z \subset V$ such that $V \backslash Z$ is disconnected.*

Proof: Let x and y be two nonadjacent nodes and Z be a minimal set of nodes such that $\langle x | Z | y \rangle$. Hence, $V \backslash Z$ is disconnected, and G can be sectioned into $G_x = (V_x, E_x)$ and $G_y = (V_y, E_y)$, where $x \in V_x$, $y \in V_y$ and $V_x \cap V_y = Z$. We show below that Z is complete.

If $|Z| = 1$, we are done. Otherwise, pick $u, v \in Z$. Because Z is minimal, there is a path from x to y with only $u \in Z$, and the same is true for v. Hence, there is a path in G_x between u and v that contains no other nodes in Z and at least one node outside Z. The same is true in G_y. Let ρ_x be the shortest such path between u and v in G_x, and ρ_y be one of the shortest such paths in G_y. Let $x' \notin Z$ be a node on ρ_x and $y' \notin Z$ be a node on ρ_y. It follows that $\{x', y'\}$ is not in G.

The cycle joining ρ_x and ρ_y has a length greater than or equal to 4. Because G is chordal, the cycle has a chord. Because $\{x', y'\}$ is not in G, $\{u, v\}$ must be a chord.   □

A node in a graph is *simplicial* if nodes adjacent to it are complete. Figure 4.11(a) has no simplicial nodes, (b) has one simplicial node a, and (c) has two simplicial nodes b and d.

Lemma 4.14 *A chordal graph G has at least one simplicial node.*[1]

Proof: Without losing generality, we assume G is connected. The lemma is true if G is complete. Assume that $G = (V, E)$ is incomplete. By Lemma 4.13 and its proof, G can be sectioned into G_0 and G_1 with the intersection $Z_1 \subset V$ complete. If $G_1 = (V_1, E_1)$ is complete, we have $z_1 \in V_1 \backslash Z_1$, which is simplicial. Otherwise, G_1 can be further sectioned into G'_1 and G_2 with the intersection $Z_2 \subset V_1$ complete. Continuing this process, eventually we obtain $G_i = (V_i, E_i)$ ($i \geq 2$), which is complete with simplicial node $z_i \in V_i \backslash Z_i$.   □

---

[1] In fact, at least two simplicial nodes exist. For the purpose of this book, however, one suffices.

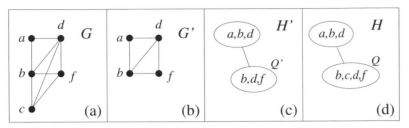

Figure 4.12: Illustration of proof for Theorem 4.10.

We are ready to prove Theorem 4.10:

Proof: [Sufficiency] We show that if G is chordal, then a JT H exists. We use induction on the number $|V|$. The statement is true when $|V| = 2$. Assume that it holds when $|V| = n$.

Suppose $|V| = n + 1$. By Lemma 4.14, a simplicial node $x \in V$ exists. Let G' be the graph obtained by removing x from G. G' is chordal because G is chordal, and x is simplicial. By the inductive assumption, a JT H' exists whose clusters are cliques of G'. Because x is simplicial in G, x and nodes adjacent to it form a clique Q. If $Q \backslash \{x\}$ is a cluster Q' in H', let H be a cluster graph obtained by replacing Q' with Q (see Figure 4.12 for an illustration of G, G', H, and H', where $x = c$). Otherwise, there exists a cluster Q' in H' such that $(Q \backslash \{x\}) \subset Q'$ (see Figure 4.13 for an illustration, where $x = e$). Let H be the cluster graph obtained by adding a cluster Q adjacent to the cluster Q'. In either case, H is a JT.

[Necessity] We show that if a JT H exists, then G is chordal. We use induction on the number $|V|$. The statement is true when $|V| = 2$. Assume that it is true when $|V| = n$.

Suppose $|V| = n + 1$. Because H is a JT, there exists a cluster Q that is adjacent to a single cluster C. Consider any $x \in Q \backslash C$. If $Q \backslash \{x\} \subset C$, we remove Q from H to get H' (see Figure 4.13 for an example). Otherwise, we replace the cluster Q by $Q' = Q \backslash \{x\}$ to get H' from H (see Figure 4.12 for an example). We also

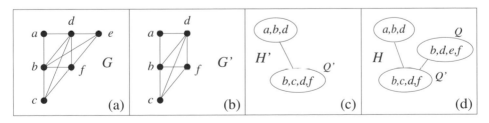

Figure 4.13: Illustration of proof for Theorem 4.10.

remove x from G to obtain G'. Because Q is the only cluster of H that contains x, H' is a JT. By the inductive assumption, G' is chordal. Because nodes adjacent to x of G are complete, x cannot be on a chordless cycle. Hence, G is also chordal.   □

Theorem 4.10 implies that, in order to obtain a JT representation of a BN, the moral graph of its DAG needs to be converted into a chordal graph. To maintain the I-mapness, links can only be *added* to the moral graph. The process of adding links to a graph to make it chordal is called *triangulation*, which is the topic of the next section.

4.6 Triangulation by Elimination

We introduce a conceptually simple operation called *node elimination* that can be used for triangulation of a graph.

Let G be an undirected graph. A node v is *eliminated* from G by adding to G links that make v simplicial and then removing v. The necessary added links are called *fill-ins*. Figure 4.14 illustrates the elimination of node x. The elimination adds the fill-in $\langle y, z \rangle$. If y is subsequently eliminated, because y is now simplicial, no fill-in needs to be added.

G is *eliminatable* if all nodes can be eliminated in sequence without any fill-ins. Nodes in Figure 4.11(c) can be eliminated in the order (b, e, c, d, f) without fill-ins; hence, the graph is eliminatable. Note that if the graph is eliminated in the order (b, c, d, e, f), a fill-in $\langle d, e \rangle$ needs to be added when eliminating c. Consequently, as long as there exists one elimination order that is fill-in free, the graph is eliminatable. On the other hand, if no such order can be found, the graph is *not* eliminatable. Figure 4.14 is not eliminatable (Exercise 6). When the graph is understood from the context, we say that a subset X of nodes is *eliminatable* if elements of X can be eliminated from the graph in some order without fill-ins.

Elimination provides a conceptually simple way to check if a graph is chordal, as established by the following theorem:

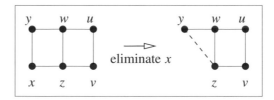

Figure 4.14: Elimination of node x. The fill-in is dashed.

Theorem 4.15 *An undirected graph G is chordal if and only if it is eliminatable.*

Proof: [Sufficiency] We show by induction that if $G = (V, E)$ is eliminatable, then it is chordal. The statement holds when $|V| = 1$, and we assume that it holds when $|V| = n$.

Suppose $|V| = n + 1$. Because G is eliminatable, there exists $v \in V$, which is simplicial. Let $G' = (V', E')$ be the graph obtained by eliminating v from G. Given that G is eliminatable and v is simplicial, G' is also eliminatable. By assumption, G' is chordal.

[Necessity] We show by induction that if G is chordal, then it is eliminatable. The statement holds when $|V| = 1$, and we assume it holds when $|V| = n$.

Suppose $|V| = n + 1$. Because G is chordal, by Lemma 4.14, G has a simplicial node v and can be eliminated without fill-ins. Let $G' = (V', E')$ be the resultant graph. Since G is chordal and v is simplicial, G' is chordal. By the inductive assumption, G' is eliminatable. ☐

Theorem 4.15 suggests the following simple algorithm to test if G is chordal:

Algorithm 4.1 (IsChordal)

Input: an undirected graph $G = (V, E)$.
Return: true if G is chordal; false otherwise.

for $i = 1$ to $|V|$, do
  search for a node v that is simplicial;
  if found, eliminate v;
  else return false;
return true;

When G is not chordal, elimination of nodes in G will have fill-ins. If these fill-ins are added back to G, the resultant graph G' will be chordal, as established by the following theorem:

Theorem 4.16 *Let $G = (V, E)$ be an undirected graph and F be a set of fill-ins produced by eliminating all nodes of G in any order. Then $G' = (V, E \cup F)$ is eliminatable.*

Proof: Let γ be the elimination order used to produce F. If nodes of G' are eliminated according to γ, each node is simplicial before it is to be eliminated. ☐

Combining Theorems 4.16 and 4.15, we have the following corollary:

Corollary 4.17 *Let $G = (V, E)$ be an undirected graph and F be a set of fill-ins produced by eliminating all nodes of G in any order. Then $G' = (V, E \cup F)$ is chordal.*

Corollary 4.17 suggests a simple modification of algorithm **IsChordal** for triangulation: At each iteration of the *for* loop, if a simplicial node cannot be found, eliminate a node and store the fill-ins. At the end of the *for* loop, return $G' = (V, E \cup F)$. This is summarized as follows:

Algorithm 4.2 (GetChordalGraph)

Input: an undirected graph $G = (V, E)$.
Return: a chordal graph $G' = (V, E')$ where $E' \supseteq E$.

$F = \emptyset$;
for $i = 1$ to $|V|$, do
    search for a node v that is simplicial;
    if found, eliminate v;
    else
        select a node w to eliminate;
        add fill-ins produced to F;
return $G' = (V, E \cup F)$;

We have studied triangulation in order to convert the moral graph of a BN into a chordal graph so that a JT representation can be constructed. Fill-ins added during triangulation destroy some graphical separation relations and hence should be kept minimal. Unfortunately, finding the chordal graph with the minimal fill-ins is NP-complete (Yannakakis [97]). One useful heuristic can be incorporated directly into the algorithm **GetChordalGraph**: When selecting a nonsimplicial node to eliminate, select the one with the minimum number of fill-ins.

4.7 Junction Trees as I-maps

Before moving into the construction of a JT from a chordal graph of a BN, we consider the issue of how good a JT representation is as an I-map. When a DAG I-map is converted to its moral graph, we lose the marginal independence among variables that may incur induced dependence. When the moral graph is converted to a chordal graph, we lose additional conditional independence by adding fill-ins. Do we incur further loss of independence when the chordal graph is converted into a JT?

The following theorem has good news and answers the question negatively. It shows that, through h-separation, a JT portrays exactly the same set of graphical separation relations as does the chordal graph, through u-separation, from which it is derived.

Theorem 4.18 *Let $G = (V, E)$ be a connected chordal graph, Ω be the set of cliques of G, and $T = (V, \Omega, F)$ be a JT. Then, for any disjoint subsets X, Y, and Z of V,*

$$\langle X | Z | Y \rangle_G \Longleftrightarrow \langle X | Z | Y \rangle_T.$$

Proof: First, we show $\langle X | Z | Y \rangle_G \Longrightarrow \langle X | Z | Y \rangle_T$. For each $x \in X$ and $y \in Y$, $\langle X | Z | Y \rangle_G$ means that $\langle x, y \rangle$ is not a link in G. Hence, x and y cannot be in the same clique in G. Denote any two clusters in T that contain x and y by Q_x and Q_y. We prove by contradiction. Suppose that on the path between Q_x and Q_y no separator is a subset of Z. Denote these separators as $S_1, \ldots, S_n$ ($n \geq 1$).

If $n = 1$, Q_x and Q_y are adjacent in T. Because $S_1 \nsubseteq Z$, we can find $s_1 \in S_1$ such that $s_1 \notin Z$. This means that both $\langle x, s1 \rangle$ and $\langle s1, y \rangle$ are links in G. Hence, $\langle X | Z | Y \rangle_G$ is false, which is a contradiction.

If $n > 1$ and no separator of S_i ($1 \leq i \leq n$) is a subset of Z, then for each i we can find $s_i \in S_i$ such that $s_i \notin Z$. Either $s_i = s_{i+1}$, in which case we have one less node to consider, or $\langle s_i, s_{i+1} \rangle$ is a link in G because s_i and s_{i+1} are contained in the same cluster in T. We have thus found a path $\langle x, s_1, \ldots, s_n, y \rangle$ on which every node $s_i \notin Z$, which is a contradiction to $\langle X | Z | Y \rangle_G$.

Next, we show $\langle X | Z | Y \rangle_G \Longleftarrow \langle X | Z | Y \rangle_T$. Let $x \in X$ and $y \in Y$ be in clusters Q_x and Q_y of T, respectively. Suppose that on the path between Q_x and Q_y, there is a separator $S \subseteq Z$. We show that $\langle x | S | y \rangle_G$ holds; hence, so does $\langle x | Z | y \rangle_G$. Assume that $\langle x | S | y \rangle_G$ does not hold. Then there exists a path $\langle x = v_0, v_1, v_2, \ldots, v_n, v_{n+1} = y \rangle$ in G not through S. That is, $v_i \notin S$ for $1 \leq i \leq n$.

Now consider T. If v_{i-1}, v_i, and v_{i+1} are not contained in the same cluster so that v_{i-1}, v_i are in cluster Q_{i-1} and v_i, v_{i+1} are in Q_{i+1}, then on the unique path between Q_{i-1} and Q_{i+1} every separator must contain v_i. Hence, on the path from Q_x to Q_y, every separator contains at least one $v_i \notin S$. This contradicts the assumption, that S is a separator between Q_x and Q_y. $\square$

Figure 4.15 shows the moral graph G (left) of the circuit in Figure 2.3, which is chordal, and its JT T (right). They are equivalent I-maps. That is, any conditional independence portrayed by G is also portrayed by T. For example, $I(a, e, b)$ can be read of G because $\langle a \, | e | \, b \rangle_G$. Given that $\{e\}$ is the separator between C_0 and C_1, we have $\langle a | e | b \rangle_T$.

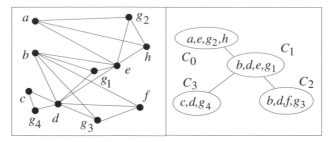

Figure 4.15: The moral graph and its JT for the circuit in Figure 2.3.

4.8 Junction Tree Construction

Once the moral graph $G = (V, E')$ of a BN is triangulated, in order to convert it into a JT representation, the cliques of the chordal graph need to be identified. From the set Ω of cliques, a junction graph $H = (V, \Omega, E)$ is defined. By removing some separators from E, a JT T can be obtained.

4.8.1 Identifying Cliques

The following proposition shows that, given a chordal graph G, if the adjacency of each node is saved just before it is eliminated, the resulting record will contain all the cliques of G.

Proposition 4.19 *Let G be a chordal graph eliminatable in the order $\gamma = (v_1, \ldots v_n)$ and Q be a clique in G. Let $C_i = \{v_i\} \cup adj(v_i) \, (1 \le i \le n)$, where $adj(v_i)$ is the adjacency of v_i when it is eliminated in γ.*
Then there exists j $(1 \le j \le n)$ such that $Q = C_j$.

Proof: Let v_k be the node with the lowest index in Q. Then $C_k \supseteq Q$. On the other hand, because G is eliminatable in γ, no fill-ins are added; hence, $Q \supseteq C_k$. $\qquad\square$

Based on Proposition 4.19, cliques in G can be efficiently identified by removing each C_i contained in some C_j. What remains is the set Ω.

4.8.2 Constructing Junction Trees

Let $H = (V, \Omega, E)$ be a junction graph, where

$$E = \{\langle Q_1, Q_2 \rangle | Q_1, Q_2 \in \Omega, Q_1 \cap Q_2 \ne \emptyset\},$$

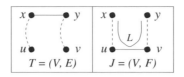

Figure 4.16: Illustration of Lemma 4.20. Each dashed line represents a number of links.

and each link $\langle Q_1, Q_2 \rangle$ is associated with a separator $S = Q_1 \cap Q_2$. Without confusion, we will refer to a link in H and its separator interchangeably. The *weight* of a separator S is its cardinality $w(S) = |S|$. The *weight* of H is the weight sum of all separators $w(H) = \sum_S w(S)$. The weight of a cluster graph is similarly defined. A cluster tree $T = (V, \Omega, E')$ of H, where $E' \subset E$ is a *maximal* cluster tree if for any cluster tree R of H we have $w(T) \geq w(R)$.

In the following, we show that, given a junction graph, a JT can be obtained by computing a maximal cluster tree.

Lemma 4.20 *Let $T = (V, E)$ and $J = (V, F)$ be two trees and $\langle x, y \rangle$ be a link in $E \backslash F$. Then there exists a link $\langle u, v \rangle \in F \backslash E$ on the path between x and y in J such that the path between u and v in T contains $\langle x, y \rangle$.*

Proof: See Figure 4.16. Because J is a tree, the path L between x and y in J is unique. Not all links on L are in T, for otherwise a cycle would be formed in T. Hence, there exist links on L that are in $F \backslash E$.

We prove the existence of $\langle u, v \rangle$ by contradiction. Suppose the link $\langle u, v \rangle$ as stated in the lemma does not exist. Then for each $\langle u, v \rangle$ that is on L and in $F \backslash E$, adding $\langle u, v \rangle$ to T forms a unique cycle that does not contain $\langle x, y \rangle$. Because deleting $\langle x, y \rangle$ from T will split T into two subtrees, it implies that each cycle involves only nodes in one of the subtrees.

We delete $\langle x, y \rangle$ from T to obtain T' and add all links that are on L and in $F \backslash E$ to T' to obtain T^*. Nodes x and y are disconnected in T'. By the argument above, they are also disconnected in T^*. Because T^* contains all links in L, L is not a path between x and y in J, which is a contradiction. $\qquad\square$

Theorem 4.21 *If a junction graph H has a junction tree, then any maximal cluster tree of H is a junction tree.*

Proof: Let $T = (V, \Omega, E)$ be a maximal cluster tree of H. Let $J = (V, \Omega, F)$ be a JT of H that has the maximal separator intersection with T. That is, for any JT

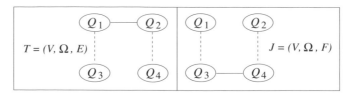

Figure 4.17: Illustration of proof for Theorem 4.21. Each dashed line represents a number of clusters.

$J' = (V, \Omega, F')$, $F \cap E \subseteq F' \cap E$ implies $F \cap E = F' \cap E$. We prove $T = J$ by contradiction.

Suppose $T \neq J$. Then there exists a separator $\langle Q_1, Q_2 \rangle \in E \setminus F$ (see Figure 4.17). According to Lemma 4.20, there exists a separator $\langle Q_3, Q_4 \rangle \in F \setminus E$ on the unique path between Q_1 and Q_2 in J such that the path between Q_3 and Q_4 in T contains $\langle Q_1, Q_2 \rangle$. Because J is a JT, we have $Q_1 \cap Q_2 \subseteq Q_3 \cap Q_4$.

We replace $\langle Q_1, Q_2 \rangle$ by $\langle Q_3, Q_4 \rangle$ in T to obtain T', which is a cluster tree. Its weight is $w(T') = w(T) + |Q_3 \cap Q_4| - |Q_1 \cap Q_2|$. Because T is maximal, $w(T) - w(T') = |Q_1 \cap Q_2| - |Q_3 \cap Q_4| \geq 0$. From $Q_1 \cap Q_2 \subseteq Q_3 \cap Q_4$, it follows $|Q_3 \cap Q_4| - |Q_1 \cap Q_2| \geq 0$. Hence, $Q_3 \cap Q_4 = Q_1 \cap Q_2$.

Next, we replace $\langle Q_3, Q_4 \rangle$ by $\langle Q_1, Q_2 \rangle$ in J to obtain J', which is a cluster tree. For any clusters Q_5 and Q_6 in J', if the path between them does not contain $\langle Q_1, Q_2 \rangle$, then it is identical to that of J. If the path contains $\langle Q_1, Q_2 \rangle$, then $Q_5 \cap Q_6 \subseteq Q_3 \cap Q_4$ because J is a JT. From $Q_3 \cap Q_4 = Q_1 \cap Q_2$, $Q_5 \cap Q_6 \subseteq Q_1 \cap Q_2$ follows; hence, J' is a JT. Cluster tree J' has exactly one more separator in common with T than J, namely, $\langle Q_1, Q_2 \rangle$. This contradicts the assumption that J has the maximal separator intersection with T. □

From Theorem 4.10, if H is obtained from a chordal graph, then a JT exists. Hence, Theorem 4.21 establishes that a JT of H can be constructed by computing a maximal cluster tree of H. A greedy algorithm (commonly referred to as Prim's algorithm for a maximal spanning tree) can be found in textbooks on discrete mathematics (e.g., Grimaldi [22]). The idea is simple: Start a JT with a single cluster and add remaining clusters one at a time such that the new cluster has the maximal separator weight with a cluster in the current JT.

Figure 4.18 illustrates the greedy algorithm. The four clusters C_0 through C_3 are shown in (a). Suppose that the JT starts with a single cluster C_2. The intersections of C_2 with each of C_0, C_1, and C_3 are $\emptyset$, $\{b, d\}$, and $\{d\}$, respectively. The weights of the corresponding separators will be 0, 2, and 1, respectively. Hence, C_1 is connected to C_2 as in (b). The remaining clusters C_0 and C_3 have the intersections $\emptyset$ and $\{d\}$ with C_2, respectively, and have the intersections $\{e\}$ and $\{d\}$ with C_1, respectively.

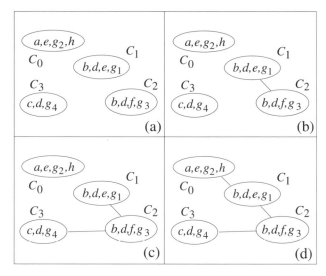

Figure 4.18: Illustration of junction tree construction.

At this point, three alternatives produce the equal total weight: connecting C_3 to C_2, connecting C_0 to C_1, or connecting C_3 to C_1. Suppose that the tie is arbitrarily broken by connecting C_3 to C_2, as in (c). Finally, the remaining cluster C_0 has the intersections $\{e\}$, $\emptyset$, and $\emptyset$ with C_1, C_2, and C_3, respectively. It is connected with C_1 as in (d). The junction tree is complete.

The preceding example demonstrates that, in general, the construction of a junction tree is *not* unique. That is, given a chordal graph, multiple corresponding junction trees exist in general. However, from Theorem 4.18, we know that, as far as the I-mapness is concerned, they are equivalent.

4.9 Bibliographical Notes

The d-separation criterion was proposed by Pearl [50]. Pearl [52] also proposed the concept of the I-map. Theorem 4.10 on the relation between the chordality of a graph and existence of a junction tree was proven by Beeri et al. [2]. Various concepts and methods relating to chordal graphs and triangulation are contained in Rose, Tarjani, and Lueker [58], Golumbic [21], Yannakakis [97], and Tarjan and Yannakakis [74].

The sufficiency for Theorem 4.18 on I-mapness of chordal graphs and junction trees, $\langle X|Z|Y\rangle_G \Longrightarrow \langle X|Z|Y\rangle_T$, is excerpted from Pearl [52], and the necessity, $\langle X|Z|Y\rangle_G \Longleftarrow \langle X|Z|Y\rangle_T$, is from Xiang [85]. Proposition 4.19 on identification of cliques was formulated by Jensen [29]. Theorem 4.21 on construction of junctions as maximal cluster trees is attributed to Jensen [28].

4.10 Exercises

1. Determine if each of the following graph separation relations holds in the DAG in Figure 4.19:

    (a) $\langle a|i|b \rangle$
    (b) $\langle a|\emptyset|c \rangle$
    (c) $\langle \{a,c\}|\{f,g\}|i \rangle$
    (d) $\langle e|\{a,b\}|g \rangle$

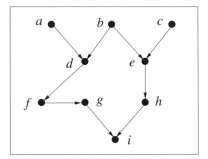

Figure 4.19: A DAG.

2. Determine if each of the following graph separation relations holds in the JT in Figure 4.20:

    (a) $\langle a|\{c,e,g\}|k \rangle$
    (b) $\langle \{i,k\}|\{e,f\}|\{a,h\} \rangle$

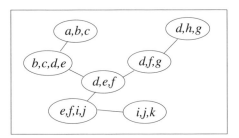

Figure 4.20: A junction tree.

3. Consider the graph G in Figure 3.12(a). Define a BN with the dependence structure G by specifying the conditional probability distribution at each node. Form $P(a,b,c,d)$ by chain rule. Test if $I(a,b,\{c,d\})$ holds by Definition 2.1.

4. Consider the DAG in Figure 4.8(a) with its moral graph in (b). The cliques in the moral graph are

$$C_0 = \{a,b\}, \quad C_1 = \{a,d\}, \quad C_2 = \{b,c\}, \quad C_3 = \{c,d,e\}.$$

Show that no cluster graph made out of these clusters is a JT.

5. The minimal I-map of a domain is given as the DAG in Figure 4.19.
   (a) Does $I(b, \{d, e\}, \{f, h\})$ hold in the domain?
   (b) Does $I(a, \emptyset, b)$ hold in the domain?
   (c) Does $I(g, \emptyset, h)$ hold in the domain?
   (d) Find a minimal undirected I-map G of the domain.
   (e) From G only, decide whether $I(a, \emptyset, b)$ holds.
6. Demonstrate that Figure 4.14 is not eliminatable.
7. Determine if the two graphs in Figure 4.21 are chordal using the algorithm **IsChordal**. If any graph is nonchordal, triangulate it using the algorithm **GetChordalGraph**.

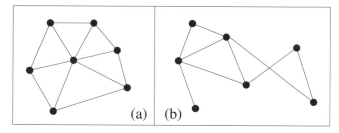

(a) (b)

Figure 4.21: Two undirected graphs.

8. Let the DAG of a BN be that in Figure 4.19. Construct a JT for the BN.

5

Belief Updating with Junction Trees

Chapter 4 discussed the conversion of the DAG structure of a BN into a junction tree. In a BN, the strength of probabilistic dependence between variables is encoded by conditional probability distributions. This quantitative knowledge is encoded in a junction tree model in terms of probability distributions over clusters. For flexibility, these distributions are often unnormalized and are termed *potentials*. This chapter addresses conversion of the conditional probability distributions of a BN into potentials in a junction tree model and how to perform belief updating by passing potentials as concise messages in a junction tree.

Section 5.2 defines basic operations over potentials: product, quotient, and marginal. Important properties of mixed operations are discussed, including associativity, order independence, and reversibility. These basic and mixed operations form the basis of message manipulation during concise message passing. Initializing of potentials in a junction tree according to the Bayesian network from which it is derived is then considered in Section 5.3. Section 5.4 presents an algorithm for message passing over a separator in a junction tree and discusses the algorithm's consequences. Extending this algorithm, Section 5.5 addresses belief updating by message passing in a junction tree model and formally establishes the correctness of the resultant belief. Processing observations is described in Section 5.6.

5.1 Guide to Chapter 5

Given a BN, its DAG structure provides the *qualitative* knowledge about the dependence among domain variables. The BN's JPD in the form of a conditional probability distribution at each node of the DAG provides *quantitative* knowledge about the strength of the dependence. Chapter 4 presented the conversion of the qualitative knowledge into a JT dependence structure. Before belief updating can be performed, the quantitative knowledge in the BN needs to be associated with

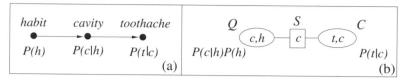

Figure 5.1: (a) The cavity BN. (b) Its JT representation.

the JT structure. Section 5.3 presents a method to assign the conditional probability distributions in the BN to the clusters in the JT. The method works as follows:

For each node v in the DAG, find a cluster in the JT such that the cluster contains both v and its parent nodes in the DAG. After the cluster is found, assign the conditional probability distribution associated with v to the cluster. As an example, consider the cavity BN shown in Figure 5.1(a) and its JT in (b). The node h has no parent, and it is contained in the cluster Q. Therefore, $P(h)$ is assigned to Q. The node c and its parent h are both contained in Q. Hence, $P(c|h)$ is also assigned to Q. Similarly, $P(t|c)$ is assigned to the cluster C. After the assignment is completed, for each cluster in the JT, the product of its assigned distributions is computed. For instance, $P(c|h)P(h)$ is computed for Q.

The product of a number of probability distributions assigned to a cluster may not be a well-defined probability distribution (e.g., $P(a|b, c)P(b)$ for some cluster $\{a, b, c\}$). In Section 3.3, we introduced the notion of *potential*, an unnormalized probability distribution that allows convenient representation of such an intermediate result. Section 5.2 presents rules for computing the product of potentials as well as rules for other operations on potentials, including quotient (similar to division) and marginal. These operations are applied to message passing over a JT representation in subsequent sections. The rules for these operations help to simplify message passing and to improve efficiency.

Section 5.4 presents the basic operation for concise message passing in a JT, which is called *absorption*. For the example in Figure 5.1, the cluster C absorbs from the cluster Q as follows: The potential $P(c|h)P(h)$ at Q is reduced (through the *marginal* operation) to a potential $P(c)$ over the separator $S = \{c\}$. The message is then sent to the cluster C, which updates its own potential to the product $P(t|c)P(c)$. Note that before the absorption, neither $P(c)$ nor $P(t)$ can be obtained at the cluster C. However, after the absorption, both of them can be computed locally at C. Several other desirable properties of absorption are also presented.

Section 5.5 considers belief updating in a general JT structure through a sequence of absorptions. Figure 5.2 demonstrates how the absorptions are performed. Each arrow illustrates an absorption and the passing of a concise message. The number beside each arrow indicates the order of the absorption. It can be seen that messages flow from C_2 and C_3 through C_1 towards C_0 and then flow from C_0 through C_1

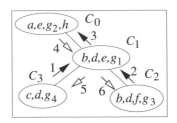

Figure 5.2: Organizing several absorptions performed over a JT.

towards C_2 and C_3. We show that after the message passing, $P(d)$ can be obtained locally from any one of C_1, C_2, and C_3, and $P(e)$ can be obtained locally from any one of C_0 and C_1. The same is true for every other variable, that is, $P(v)$ can be obtained locally from any cluster that contains v.

What we are more interested in is, say $P(d|e = e_0)$, after an observation $e = e_0$ is made. Section 5.6 presents how the observation can be entered into the JT by a simple product operation in any cluster that contains e. Then, after the absorptions shown in Figure 5.2, again, $P(d|e = e_0)$ can be obtained in any cluster that contains d. The goal of exact belief updating for BNs of nontree structures by concise message passing in a single graph structure has thus been attained.

5.2 Algebraic Properties of Potentials

Recall from Section 3.3 that a probability distribution over a cluster can be represented in terms of a *potential*. Potentials allow intermediate results during belief updating to be non-normalized and hence have computational advantage over probability distributions. Potentials will be used for belief updating in a junction tree representation of a BN. Potential algebra is introduced in this section.

Formally, a *potential* $B(X)$ over a set X of variables is a function from the domain D_X of X to the nonnegative real numbers such that for at least one $x \in D_X$, $B(x) > 0$. Subscripts may be used to differentiate potentials over the same set of variables such as $B_1(X)$ and $B_2(X)$. For discrete variables, a potential can be expressed as a table. Consider $X = \{u, v\}$ with $D_u = \{0, 1\}$ and $D_v = \{0, 1, 2\}$. A potential $B_1(X)$ is shown in Table 5.1.

Table 5.1: *A potential* $B_1(x)$

(u, v)	$B_1()$	(u, v)	$B_1()$
$(0, 0)$	0.2	$(1, 0)$	3.4
$(0, 1)$	1.2	$(1, 1)$	0
$(0, 2)$	0.9	$(1, 2)$	5.6

Table 5.2: *(Left) A potential $B_2(x)$; (Right) The product $B_3(x) = B_1(x) * B_2(x)$*

(u, v)	$B_2()$	(u, v)	$B_2()$	(u, v)	$B_3()$	(u, v)	$B_3()$
(0, 0)	0.3	(1, 0)	3.0	(0, 0)	0.06	(1, 0)	10.2
(0, 1)	2.0	(1, 1)	1.0	(0, 1)	2.4	(1, 1)	0
(0, 2)	4.0	(1, 2)	5.0	(0, 2)	3.6	(1, 2)	28

5.2.1 Product

Let $B_1(X)$ and $B_2(X)$ be potentials over X. Their *product*, denoted by $B_1(X) * B_2(X)$, is a potential $B_3(X)$ such that for each $x \in D_X$,

$$B_3(x) = B_1(x) * B_2(x).$$

For example, given $B_1(X)$ (Table 5.1) and $B_2(X)$ (Table 5.2 (left)), their product $B_3(X)$ is shown in Table 5.2 (right).

Products of potentials over distinct sets of variables are often useful. We introduce *projection* to facilitate the definition. Let x be a configuration of a set X of variables. The value of $v \in X$ in x, denoted by $v^{\downarrow x}$, is the *projection* of x onto v. Similarly, we can define the projection of x onto a subset $Z \subseteq X$. A configuration z is the *projection* of x onto Z, denoted by $z = Z^{\downarrow x}$, if for each $v \in Z$, its value in z is identical to that in x. For example, let $X = \{a, b, c, d\}$, $Z = \{a, c\}$, and $x = (a_0, b_3, c_2, d_5)$. Then, $d^{\downarrow x} = d_5$ and $Z^{\downarrow x} = (a_0, c_2)$.

We now define the potential product over distinct sets.

Definition 5.1 *Let X, Y be distinct sets of variables and $W = X \cup Y$. Let $B(X)$ and $B(Y)$ be two potentials. The **product** $B(X) * B(Y)$ is a potential $B(W)$ such that for each $w \in D_W$,*

$$B(w) = B(X^{\downarrow w}) * B(Y^{\downarrow w}).$$

Table 5.3 shows $B_4(x, y)$, $B_5(y, z)$ and $B_6(x, y, z) = B_4(x, y) * B_5(y, z)$, where all variables are binary.

The product of potentials satisfies the *commutativity* and *associativity* as follows. The proof is straightforward.

Theorem 5.2 *Let $B(X)$, $B(Y)$, and $B(Z)$ be potentials. The following hold:*

1. *Commutativity: $B(X) * B(Y) = B(Y) * B(X)$.*
2. *Associativity: $[B(X) * B(Y)] * B(Z) = B(X) * [B(Y) * B(Z)]$.*

Table 5.3: $B_6(x, y, z) = B_4(x, y) * B_5(y, z)$

(x, y)	$B_4()$	(x, y)	$B_4()$	(y, z)	$B_5()$	(y, z)	$B_5()$
(0, 0)	0.2	(1, 0)	3.4	(0, 0)	0.4	(1, 0)	2.0
(0, 1)	1.2	(1, 1)	0	(0, 1)	0.8	(1, 1)	0.1

(x, y, z)	$B_6()$	(x, y, z)	$B_6()$
(0, 0, 0)	0.08	(1, 0, 0)	1.36
(0, 0, 1)	0.16	(1, 0, 1)	2.72
(0, 1, 0)	2.4	(1, 1, 0)	0
(0, 1, 1)	0.12	(1, 1, 1)	0

5.2.2 Quotient

Next, we define the *quotient* of two potentials. To handle dividing by zero, we introduce *zero-consistency*. Let $B(X)$ and $B(Y)$ be potentials and $W = X \cup Y$. If for each configuration w of W, we have

$$B(X^{\downarrow w}) = 0,$$

whenever

$$B(Y^{\downarrow w}) = 0,$$

then $B(X)$ is said to be *zero-consistent* with $B(Y)$. The potential $B_7(x, y)$ in the Table 5.4 (left) is not zero-consistent with $B_8(y, z)$ (right), for $B_8(y = 1, z = 0) = 0$ but $B_7(x = 1, y = 1) = 1.0$; however, $B_7(x, y)$ will be zero-consistent with $B_8(y, z)$ if $B_7(x = 1, y = 1) = 0$.

Intuitively, if $B(X)$ is zero-consistent with $B(Y)$, then whenever a configuration w is impossible according to $B(Y)$, it must be impossible according to $B(X)$. Note that zero-consistency is not commutative. When $B(X)$ is zero-consistent with $B(Y)$, $B(Y)$ may not be zero-consistent with $B(X)$. For instance, the $B_8(y, z)$ above is zero-consistent with $B_7(x, y)$ even though $B_7(x, y)$ is not zero-consistent with $B_8(y, z)$.

Table 5.4: *(Left) Potentials $B_7(x, y)$; (Right) $B_8(x, z)$*

(x, y)	$B_7()$	(x, y)	$B_7()$	(y, z)	$B_8()$	(y, z)	$B_8()$
(0, 0)	0.3	(1, 0)	2.0	(0, 0)	2.5	(1, 0)	0
(0, 1)	0	(1, 1)	1.0	(0, 1)	0.8	(1, 1)	0

Table 5.5: *Potential $B_9(x, y, z)$*

(x, y, z)	$B_9()$	(x, y, z)	$B_9()$
$(0, 0, 0)$	0.12	$(1, 0, 0)$	0.8
$(0, 0, 1)$	0.375	$(1, 0, 1)$	2.5
$(0, 1, 0)$	0	$(1, 1, 0)$	0
$(0, 1, 1)$	0	$(1, 1, 1)$	0

Definition 5.3 *Let $B(X)$ be zero-consistent with $B(Y)$ and $W = X \cup Y$. The* **quotient** *of $B(X)$ over (divided by) $B(Y)$, denoted by $B(X)/B(Y)$, is a potential $B(W)$ such that for each $w \in D_W$,*

$$B(w) = \begin{cases} B(X^{\downarrow w})/B(Y^{\downarrow w}) & \text{if } B(Y^{\downarrow w}) > 0, \\ 0 & \text{if } B(Y^{\downarrow w}) = 0. \end{cases}$$

$B(X)$ is referred to as the **numerator** *and $B(Y)$ as the* **denominator**.

Intuitively, whenever $B(Y^{\downarrow w}) = 0$, w is impossible according to both $B(X)$ and $B(Y)$, and hence $B(w) = 0$. As an example, we replace $B_7(x = 1, y = 1) = 0$ in Table 5.4. The quotient $B_9(x, y, z) = B_7(x, y)/B_8(y, z)$ is shown in Table 5.5.

The following theorem establishes the *associativity* for the product and quotient of potentials.

Theorem 5.4 *The product and quotient of potentials are* **associative**. *That is, if $B(Y)$ is zero-consistent with $B(Z)$, then*

$$[B(X) * B(Y)]/B(Z) = B(X) * [B(Y)/B(Z)].$$

Proof: Let w be a configuration of $W = X \cup Y \cup Z$, $x = X^{\downarrow w}$, $y = Y^{\downarrow w}$, and $z = Z^{\downarrow w}$. If $B(z) > 0$, then

$$(B(x) * B(y))/B(z) = B(x) * (B(y)/B(z)).$$

If $B(Y)$ is zero-consistent with $B(Z)$, then $B(X) * B(Y)$ is zero-consistent with $B(Z)$. Hence, when $B(z) = 0$, each term in the equation is zero. □

The order of the product and quotient can be altered as shown below:

Theorem 5.5 *The product and quotient of potentials are* **order independent**. *That is, if $B(X)$ is zero-consistent with $B(Z)$, then*

$$[B(X) * B(Y)]/B(Z) = [B(X)/B(Z)] * B(Y).$$

Proof:

$(B(X) * B(Y))/B(Z)$
$= (B(Y) * B(X))/B(Z)$ (commutativity of product)
$= B(Y) * (B(X)/B(Z))$ (associativity of product and quotient)
$= (B(X)/B(Z)) * B(Y)$ (commutativity of product)   □

The following theorem establishes the reversibility of potentials.

Theorem 5.6 *The product and quotient of potentials are **reversible**. That is, if $B(X)$ is zero-consistent with $B(Y)$, then*

$$[B(X) * B(Y)]/B(Y) = [B(X)/B(Y)] * B(Y) = B(X).$$

Proof: Let w be a configuration of $W = X \cup Y$, $x = X^{\downarrow w}$, and $y = Y^{\downarrow w}$. If $B(y) > 0$, then

$$[B(x) * B(y)]/B(y) = [B(x)/B(y)] * B(y) = B(x).$$

When $B(y) = 0$, because $B(X)$ is zero-consistent with $B(Y)$, we have $B(x) = 0$. From the definition of a quotient, each term in the equation is zero.   □

In summary, product and quotient operations on potentials can be performed in the same way as the corresponding operations on real numbers if care is taken with respect to zero-consistency.

5.2.3 Marginal

Marginals of potentials are similar to marginal distributions.

Definition 5.7 *Let $B(X)$ be a potential and $Y \subset X$. A potential $B(Y)$ is a **marginal** of $B(X)$ if for each $y \in D_Y$,*

$$B(y) = \sum_{x:Y^{\downarrow x}=y} B(x).$$

That is, $B(y)$ is obtained by summing $B(x)$ for each x such that the projection of x onto Y equals y. Denote the marginal by $B(Y) = \sum_{X \backslash Y} B(X)$.

Table 5.6 shows a marginal $B_{10}(x, y) = \sum_z B_6(x, y, z)$.

The following theorem provides a useful rule to simplify the computation of marginals.

Table 5.6: $B_{10}(x, y) = \sum_z B_6(x, y, z)$

(x, y)	$B_{10}()$	(x, y)	$B_{10}()$
$(0, 0)$	0.24	$(1, 0)$	4.08
$(0, 1)$	2.52	$(1, 1)$	0

Theorem 5.8 *Let $B(X)$ and $B(Y)$ be potentials and $Z \subseteq Y \backslash X$. Then*

$$\sum_Z [B(X) * B(Y)] = B(X) * \sum_Z B(Y).$$

Proof: See Figure 5.3, which illustrates a set $W = X \cup Y$ and its subsets $Y = U \cup Z$ and $V = X \cup U$, where $U \cap Z = \emptyset$ and $X \cap U = \emptyset$. The figure also shows a configuration $w \in D_W$ and its projections $x = X^{\downarrow w}$, $y = Y^{\downarrow w}$, $v = V^{\downarrow w}$ and $u = U^{\downarrow w}$.

The term on the left of the equation defines a potential $B(V)$ such that for each $v \in D_V$,

$$B(v) = \sum_{w:V^{\downarrow w}=v} B(X^{\downarrow w}) * B(Y^{\downarrow w}).$$

The term on the right defines a potential $B'(V)$ such that for each $v \in D_V$ and $u = U^{\downarrow v}$

$$B'(v) = B(X^{\downarrow v}) * \sum_{y:U^{\downarrow y}=u} B(y).$$

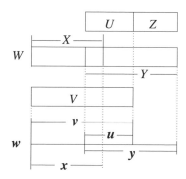

Figure 5.3: Illustration for proof of Theorem 5.8.

Because for each w such that $V^{\downarrow w} = v$,

$$X^{\downarrow w} = X^{\downarrow (V^{\downarrow w})} = X^{\downarrow v},$$

we have

$$B(v) = \sum_{w:V^{\downarrow w}=v} B(X^{\downarrow v}) * B(Y^{\downarrow w}) = B(X^{\downarrow v}) * \sum_{w:V^{\downarrow w}=v} B(Y^{\downarrow w}).$$

Consider a given $v \in D_V$ and $u = U^{\downarrow v}$, as shown in Figure 5.3. For each w such that $V^{\downarrow w} = v$, there exists a unique y such that $Y^{\downarrow w} = y$ and $U^{\downarrow y} = u$. On the other hand, for each y such that $U^{\downarrow y} = u$, a unique w exists such that $Y^{\downarrow w} = y$ and $V^{\downarrow w} = v$. Hence,

$$\sum_{w:V^{\downarrow w}=v} B(Y^{\downarrow w}) = \sum_{y:U^{\downarrow y}=u} B(y),$$

which implies $B(v) = B'(v)$.   □

Consider $\sum_z B_4(x, y) * B_5(y, z)$. Computing $B_6(x, y, z)$ first, as in Table 5.3, and then $\sum_z B_6(x, y, z)$, as in Table 5.6, takes eight multiplications and four additions. Computing $\sum_z B_5(y, z) = B'(y) = (1.2, 2.1)$ first and then $B_4(x, y) * B'(y)$ takes two additions and four multiplications.

5.3 Potential Assignment in Junction Trees

To use a JT representation of a BN for belief updating, the JPD of the BN needs to be converted into the belief over the JT. As in a BN, we seek a concise and localized belief representation by exploring the independence encoded in the JT.

Let T be a JT over a set V of variables. Assign each cluster Q in T a potential $B_Q(Q)$ and each separator S a potential $B_S(S)$. Let $\mathcal{B}$ denote the set of all such potentials. We will call $\mathcal{T} = (V, T, \mathcal{B})$ a *junction tree representation*[1] of the domain V. When there is no confusion, we simply refer to $\mathcal{T}$ as a JT.

Section 3.5 defines the notion of *consistence* between two adjacent clusters Q and C. Now that a potential is associated with their separator S, the notion needs to be extended to include a constraint for the separator potential. The concept of *local consistence* of a cluster tree representation introduced in Section 3.7 needs to be extended accordingly. These concepts are extended here in. For readers' convenience, the notion of *global consistence* defined in Section 3.7 is repeated here.

---

[1] Although the term *junction tree representation* has been used informally so far, this is how the term is formally defined.

Definition 5.9 *Let* $T = (V, T, \mathcal{B})$ *be a cluster tree representation. Let Q and C be two adjacent clusters with the separator S and their associated potentials be $B_Q(Q)$, $B_C(C)$, and $B_S(S)$, respectively. Clusters Q and C are said to be* **consistent** *if*

$$\sum_{Q\setminus S} B_Q(Q) = \text{const}_1 * B_S(S) = \text{const}_2 * \sum_{C\setminus S} B_C(C).$$

T *is said to be* **locally consistent** *if every pair of adjacent clusters in T are consistent. T is said to be* **globally consistent** *if for any two clusters Q and C (not necessarily adjacent), it holds that*

$$\sum_{Q\setminus C} B_Q(Q) = \text{const} * \sum_{C\setminus Q} B_C(C).$$

The following theorem establishes the relation between the JPD of a domain and the cluster and separator potentials of a JT representation. We have written $Q \in T$ in the theorem instead of $Q \in \Omega$ for convenience and used a similar notation $S \in T$ for a separator S.

Theorem 5.10 *Let $T = (V, T, \mathcal{B})$ be a JT representation such that T is an I-map over V. For each cluster Q, $B_Q(Q) = \text{const}_1 * P(Q)$, and for each separator S, $B_S(S) = \text{const}_2 * P(S)$. Then the joint probability distribution over V is*

$$P(V) = \text{const} * \prod_{Q\in T} B_Q(Q) \Big/ \prod_{S\in T} B_S(S).$$

Proof: Without losing generality, assume $\text{const}_1 = \text{const}_2 = 1$. We then only have to show

$$P(V) = \prod_{Q\in T} P(Q) \Big/ \prod_{S\in T} P(S).$$

We prove this by induction on the number n of clusters. When $n = 2$, let the two clusters be C and Q. We have

$$P(V) = P(C)P(V\setminus C|C) = P(C)P(Q\setminus S|C) = P(C)P(Q\setminus S|S)$$
$$= P(C)P(Q)/P(S),$$

where the third equation is due to the I-mapness of T.

Assume that the statement is true when $n = k \geq 2$. Consider $n = k + 1$. Pick a terminal cluster Q with separator R. Define a JT representation $(V', T', \mathcal{B}')$, where $V' = V\setminus(Q\setminus R)$, T' is obtained by removing Q and R from T, and $\mathcal{B}'$ is the set of potentials associated with T'. By the inductive assumption,

$$P(V') = \prod_{Q \in T'} P(Q) \Big/ \prod_{S \in T'} P(S).$$

Because T is an I-map, we have

$$P(V) = P(V')P(Q \backslash R | R) = P(V')P(Q)/P(R) = \prod_{Q \in T} P(Q) \Big/ \prod_{S \in T} P(S).$$

$\square$

On the basis of Theorem 5.10, we define the *system potential* for T as

$$B_T(V) = \prod_{Q \in T} B_Q(Q) \Big/ \prod_{S \in T} B_S(S). \tag{5.1}$$

Next, we consider how to assign local potentials in a JT representation so that its system potential is equivalent to the JPD of the corresponding Bayesian network $(V, G, \mathcal{P})$. One simple method is introduced below. Given a node v in G with its parents $\pi(v)$, we refer to $\{v\} \cup \pi(v)$ as the *family* of v denoted by *fmly*(v).

Let a BN be $(V, G, \mathcal{P})$ and a JT obtained from G be $T = (V, \Omega, E)$. For each cluster Q and each separator S in T, associate it with a *uniform* potential. A potential $B(Q)$ is *uniform* if for each $q \in D_Q$, $B(q) = 1$. For each node v in G, find a cluster Q_v in T such that *fmly*$(v) \subseteq Q_v$ and break ties arbitrarily. Update $B(Q_v)$ to the product $B(Q_v) * P(v|\pi(v))$. As an example, consider the BN in Figure 2.6 and its JT in Figure 4.7. Figure 5.4 illustrates the potential of each cluster.

Clearly, the preceding potential assignment satisfies

$$B_T(V) = \prod_{v \in V} P(v|\pi(v)) = P(V). \tag{5.2}$$

That is, the system potential is equal to the JPD defined by the BN. Hence, a JT representation has been obtained that is equivalent (in the sense of JPD) to the original BN.

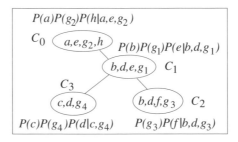

Figure 5.4: The potential of each cluster in the JT of Figure 4.7.

However, the JT representation is not locally consistent (and hence not globally consistent by Theorem 3.7). In Figure 5.4, only the potential of one cluster C_3 satisfies $B_{C_3}(C_3) = P(C_3)$ because

$$P(c)P(g_4)P(d|c, g_4) = P(c, d, g_4).$$

The potential of any other cluster is not equal to the probability distribution over the variables of the cluster. Hence, for some variables (e.g., for the variable f), their marginal distributions cannot be computed from the potential of an arbitrary containing cluster. In the next section we present a method for concise message passing in a JT that will bring the JT to consistence and therefore allow the computation of a variable marginal locally from any containing cluster.

5.4 Passing Belief over Separators

We present a method developed by Jensen et al. [30] for belief updating in a JT representation of a BN through concise message passing. The method is based on a primitive operation called *absorption*, which passes concise messages between two adjacent clusters in a JT $\mathcal{T}$.

Algorithm 5.1 (Absorption) *Let C and Q be adjacent clusters with separator S in a JT, and $B_C(C)$, $B_Q(Q)$, $B_S(S)$ be their potentials such that $B_Q(Q)$ is zero-consistent with $B_S(S)$. Then C **absorbs** from Q by performing the following:*

1. Updating $B_S(S)$ to $B'_S(S) = \sum_{Q \backslash S} B_Q(Q)$.
2. Updating $B_C(C)$ to $B'_C(C) = B_C(C) * \frac{B'_S(S)}{B_S(S)}$.

Absorption is well defined because $B'_S(S)$ is zero-consistent with $B_S(S)$. **Absorption** is one form of message passing. When C absorbs from Q, messages flow from Q to S and then to C.

To illustrate **Absorption**, consider the cavity example from Section 2.5. Figure 5.5 and Table 5.7 show its JT representation with potential assignment, where 1 denotes a uniform potential. After C absorbs from Q, the new potentials are shown in Table 5.8.

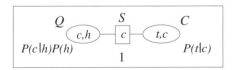

Figure 5.5: Junction representation of the cavity BN.

Table 5.7: *Potential assignment to the cavity JT*

(c, h)	$B_Q()$	c	$B_S()$	(t, c)	$B_C()$
(yes, good)	0.07	yes	1.0	(yes, yes)	0.85
(yes, bad)	0.24	no	1.0	(yes, no)	0.05
(no, good)	0.63			(no, yes)	0.15
(no, bad)	0.06			(no, no)	0.95

From Step 1 of Algorithm 5.1, it is clear that the separator S and the cluster Q are consistent after absorption. On the other hand, C may not be consistent with S yet and hence may not be consistent with Q. To bring C and Q into consistence, another absorption of Q from C is necessary, as will be seen. The *supportiveness* of the separator is introduced next to ensure that absorption can be performed in both directions.

A separator S is said to be *supportive* if absorption can be performed in *either* direction (C can absorb from Q and vice versa). A JT representation $\mathcal{T}$ is *supportive* if every separator is supportive. Clearly, if the separator potentials are assigned according to the method in Section 5.3, then $\mathcal{T}$ is supportive. Although this initial supportiveness ensures that an absorption can be performed over each separator in any direction, after the absorption, the supportiveness may change and prevent the performance of the next absorption. Lemma 5.11 shows that this will not happen and that the supportiveness of a JT is invariant under absorption.

Lemma 5.11 *Let the separator S between two clusters C and Q be supportive. Then S is supportive after C absorbs from Q.*

Proof: It suffices to show that $B'_C(C)$ is zero-consistent with $B'_S(S)$. Because

$$B'_C(C) = B_C(C) * \frac{B'_S(S)}{B_S(S)},$$

this is clearly true. □

Table 5.8: *Potentials after C absorbs from S*

(c, h)	$B_Q()$	c	$B_S()$	(t, c)	$B_C()$
(yes, good)	0.07	yes	0.31	(yes, yes)	0.2635
(yes, bad)	0.24	no	0.69	(yes, no)	0.0345
(no, good)	0.63			(no, yes)	0.0465
(no, bad)	0.06			(no, no)	0.6555

From Eq. (5.2), the initial system potential of the JT is equivalent to the JPD of the BN. It is important that the absorption does not alter the system potential while bringing the JT into consistence. Lemma 5.12 shows that this is indeed the case and that the system potential is invariant under absorption.

Lemma 5.12 *Let $\mathcal{T}$ be a supportive JT. Then $B_T(V)$ is invariant after an absorption.*

Proof: Consider clusters C and Q and their separator S. Suppose C absorbs from Q. **Absorption** updates $B_C(C)$ and $B_S(S)$ into $B'_C(C)$ and $B'_S(S)$. According to Eq. (5.1), it suffices to show

$$\frac{B'_C(C)}{B'_S(S)} = \frac{B_C(C)}{B_S(S)}$$

as follows:

$$
\begin{aligned}
B'_C(C)/B'_S(S) &= [B_C(C) * B'_S(S)/B_S(S)]/B'_S(S) && \text{(Algorithm 5.1)} \\
&= [B_C(C)/B_S(S)] * B'_S(S)/B'_S(S) && \text{(Theorem 5.5)} \\
&= B_C(C)/B_S(S) && \text{(Theorem 5.6)}
\end{aligned}
$$

$\square$

The purpose of **Absorption** is to bring adjacent clusters in a JT representation into consistence. Proposition 5.13 shows that this effect is indeed achieved. That is, two clusters are *consistent* after an absorption is performed over the separator at each direction.

Proposition 5.13 *Let the separator S between two clusters C and Q be supportive. Clusters C and Q are consistent if C absorbs from Q followed by absorption of Q from C.*

The result holds even if the potential in C changes between the two absorptions as long as S remains supportive.

Proof: After C absorbs from Q, denote updated potentials as follows:

$$B'_C(C) \qquad B'_S(S) = \sum_{Q \setminus S} B_Q(Q) \qquad B'_Q(Q) = B_Q(Q)$$

After Q absorbs from C, denote updated potentials as follows:

$$B^*_C(C) = B'_C(C) \qquad B^*_S(S) = \sum_{C \setminus S} B'_C(C) \qquad B^*_Q(Q) = B_Q(Q) * B^*_S(S)/B'_S(S)$$

Because

$$\sum_{Q \setminus S} B_Q^*(Q) = \sum_{Q \setminus S} B_Q(Q) * B_S^*(S)/B_S'(S)$$

$$= [B_S^*(S)/B_S'(S)] * \sum_{Q \setminus S} B_Q(Q) \quad \text{(Theorems 5.4 and 5.8)}$$

$$= [B_S^*(S)/B_S'(S)] * B_S'(S)$$

$$= B_S^*(S), \qquad\qquad\qquad \text{(Theorem 5.6)}$$

C and Q are consistent. Inasmuch as the proof depends on $B_C'(C)$ but not $B_C(C)$, the result holds even if $B_C'(C)$ changes after the first absorption. $\qquad\square$

When C absorbs from Q, each value in $B_Q(Q)$ is summed, and each value in $B_C(C)$ is updated. Hence, the complexity of absorption is exponential on $\max(|Q|, |C|)$. In other words, the complexity is $O(2^{\max(|Q|,|C|)})$.

5.5 Passing Belief through a Junction Tree

Proposition 5.13 suggests that, to make $\mathcal{T}$ locally consistent, absorptions need to be performed along every separator in both directions. These absorptions can be organized into two rounds of concise message passing, as illustrated in Figure 5.6:

To start with, select a cluster arbitrarily, say, C_0, and direct links of $\mathcal{T}$ away from C (white arrows). Now C_0 becomes the root of the junction tree and C_2 and C_3 become leaves. Next, C_1 absorbs from C_2 and from C_3 followed by absorption of C_0 from C_1. This is an *inward* message passing from leaves towards the root (along black arrows). Finally, C_1 absorbs from C_0 followed by absorption of C_2 from C_1 and absorption of C_3 from C_1. This is an *outward* message passing from the root towards leaves (shown by white arrows).

The message passing can generally be described in an object-oriented fashion as follows. The outward message passing is specified in **DistributeEvidence**.

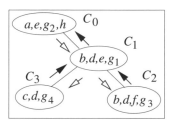

Figure 5.6: Illustration of **CollectEvidence** and **DistributeEvidence**.

Algorithm 5.2 (DistributeEvidence) *Let Q be a cluster in $\mathcal{T}$. A **caller** is either an adjacent cluster or $\mathcal{T}$. When the caller calls in Q, it does the following:*

1. *If caller is a cluster, Q absorbs from it.*
2. *Cluster Q calls **DistributeEvidence** in each adjacent cluster except caller.*

In Figure 5.6, $\mathcal{T}$ calls C_0 to start **DistributeEvidence**. In response, C_0 calls C_1. Cluster C_1 absorbs from C_0 and then calls C_2 and C_3. After C_2 (C_3) absorbs from C_1, it terminates. The inward message passing is specified in **CollectEvidence**.

Algorithm 5.3 (CollectEvidence) *Let Q be a cluster in $\mathcal{T}$. A **caller** is either an adjacent cluster or $\mathcal{T}$. When the caller calls in Q, it does the following:*

1. *Cluster Q calls **CollectEvidence** in each adjacent cluster except caller.*
2. *After each called cluster has finished, Q absorbs from it.*

In Figure 5.6, $\mathcal{T}$ calls C_0 to start **CollectEvidence**. In response, C_0 calls C_1, which in turn calls C_2 and C_3. When C_2 (C_3) is called, it has no adjacent cluster except the caller and hence returns immediately, causing C_1 to absorb from C_2 (C_3). After C_1 finishes, C_0 absorbs from C_1 and terminates **CollectEvidence**. The two rounds of message passing are combined in **UnifyBelief**.

Algorithm 5.4 (UnifyBelief) $\mathcal{T}$ *selects a cluster Q and calls **CollectEvidence** in Q. After it finishes, $\mathcal{T}$ calls **DistributeEvidence** in Q.*

The following theorem establishes that **UnifyBelief** renders $\mathcal{T}$ locally consistent.

Theorem 5.14 *Let $\mathcal{T}$ be supportive. Then after **UnifyBelief**, $\mathcal{T}$ is locally consistent.*

Proof: We prove the theorem by induction on the number k of clusters in $\mathcal{T}$. The theorem is trivially true for $k = 1$. Assume that it is true for $k = n \geq 1$, and consider $k = n + 1$. Suppose **UnifyBelief** is performed in $\mathcal{T}$ with a root Q^*. Let $Q \neq Q^*$ be a cluster with a single adjacent cluster C. Such a Q exists because $k > 1$ and $\mathcal{T}$ is a tree. Let $\mathcal{T}'$ be the JT with Q removed from $\mathcal{T}$. Because Q is a leaf in $\mathcal{T}$, **UnifyBelief** can be viewed as follows:

1. C absorbs from Q.
2. $\mathcal{T}'$ calls **UnifyBelief** in Q^*.
3. Q absorbs from C.

Because $k = n$ for T', T' is locally consistent after the second step by assumption. According to Proposition 5.13, Q and C are consistent after the third step. $\square$

From Theorems 5.14 and 3.7, we derive the following result on the global consistency.

Corollary 5.15 *Let* T *be supportive. Then after* **UnifyBelief**, T *is globally consistent.*

Theorem 5.16 tells us that the marginal distribution of any variable can be computed from any cluster in T. In fact, the marginal over any subset of any cluster can be computed effectively using the cluster potential.

Theorem 5.16 *Let* $T = (T, B_T(V))$ *be a locally consistent JT representation. Then for each cluster X,*

$$B_X(X) = \sum_{V \setminus X} B_T(V).$$

Proof: We prove by induction on the number k of clusters in T. The statement holds if $k = 1$. Assume that it holds when $k = n \geq 1$. Consider $k = n + 1$. Let Q be a cluster with a single adjacent cluster C and their separator be S.

Let T' be a JT obtained by removing Q and S from T, and $V' = V \setminus (Q \setminus S)$. Then $T' = (T', B_{T'}(V'))$. From Eq. (5.1), it holds that

$$B_T(V) = B_{T'}(V') * B_Q(Q) / B_S(S). \tag{5.3}$$

Because T is locally consistent, we have

$$\sum_{Q \setminus S} B_Q(Q) = B_S(S) = \sum_{C \setminus S} B_C(C), \tag{5.4}$$

where we have assumed a unit value of all the constants for simplicity. Furthermore,

$$
\begin{aligned}
\sum_{Q \setminus S} B_T(V) &= \sum_{Q \setminus S} B_{T'}(V') * B_Q(Q) / B_S(S) &&\text{(Eq. (5.3))} \\
&= [B_{T'}(V') / B_S(S)] * \sum_{Q \setminus S} B_Q(Q) &&\text{(Theorems 5.4, 5.2 and 5.8)} \\
&= [B_{T'}(V') / B_S(S)] * B_S(S) &&\text{(Eq. (5.4))} \\
&= B_{T'}(V'). &&\text{(Theorem 5.6)}
\end{aligned}
\tag{5.5}
$$

Hence for every cluster X in $\mathcal{T}'$, we have

$$\sum_{V \setminus X} B_T(V) = \sum_{V' \setminus X} \sum_{Q \setminus S} B_T(V)$$

$$= \sum_{V' \setminus X} B_{T'}(V') \qquad \text{(Eq. (5.5))}$$

$$= B_X(X). \qquad \text{(inductive assumption)}$$

For $X = Q$,

$$\sum_{V \setminus Q} B_T(V) = \sum_{V' \setminus S} B_{T'}(V') * B_Q(Q)/B_S(S) \qquad \text{(Eq. (5.3))}$$

$$= [B_Q(Q)/B_S(S)] * \sum_{V' \setminus S} B_{T'}(V') \qquad \text{(Theorem 5.8)}$$

By recursively marginalizing out variables contained in each terminal cluster of $\mathcal{T}'$, we have

$$\sum_{V' \setminus S} B_{T'}(V') = \sum_{C \setminus S} B_C(C).$$

Using Eq. (5.4) and Theorem 5.6, we obtain

$$\sum_{V \setminus Q} B_T(V) = [B_Q(Q)/B_S(S)] * B_S(S) = B_Q(Q).$$

$\square$

Let n be the number of clusters in $\mathcal{T}$ and q be the cardinality of the largest cluster. **UnifyBelief** performs two absorptions over each separator. Hence, its complexity is linear on n (there are $n - 1$ separators) and exponential on q. Using the $O()$ notation, we state the complexity as $O(n\,2^q)$. This implies that **UnifyBelief** is efficient as long as the largest cluster is reasonably small.

The value of q depends on both the connectivity of the original DAG dependence structure as well as the conversion process from the DAG to the JT structure. The connectivity of the DAG affects q in two ways: Because for each variable v, its family $fmly(v)$ is completed during the moralization we have

$$q \geq \max_{v \in V} |fmly(v)|.$$

That is, q is lower bounded by the size of the largest family in the DAG. Furthermore, the number of cycles and how they are connected also affect the value of q.

Next, consider the conversion process. The moralization step in the conversion process is deterministic. Once the chordal structure is obtained, the number of clusters in the JT and their cardinalities are fixed. The only step that can make a difference is then the triangulation, where as few as possible fill-ins should be added. Although heuristics can be used to keep the fill-ins minimal in many cases, the general problem of finding the chordal graph with the minimal fill-ins is NP-complete, as discussed in Section 4.6.

To keep the value q small, another opportunity that can be explored is to keep the original DAG structure sparse by not including arcs that correspond to the weak dependence. Loosely speaking, the dependence between a parent variable y and a child variable x is *weak* if $P(x|y)$ is close to $P(x)$. In other words, y does not have a strong influence on x, or, alternatively, knowing the value of y does not change one's belief in x significantly.

5.6 Processing Observations

When the values of a subset X of variables are observed, it is desirable to compute $P(v|X = x)$ for some $v \in V$. From Sections 2.3 and 3.3,

$$P(V|x) = P(V, x)/P(x)$$

can be computed by setting $P(v) = 0$ for each configuration v of V such that $X^{\downarrow v} \neq x$ and then *normalizing* the resultant potential. Formally, we define an *observation* on $x \in X$ as a function

$$obs(x) = \begin{cases} 1 & \text{if } x = x^{\downarrow x} \\ 0 & \text{otherwise.} \end{cases}$$

We have

$$P(V|x) * P(x) = P(V) * \prod_{x \in X} obs(x).$$

Under the JT representation (Eq. (5.1)), this becomes

$$P(V|x) * P(x) = B_T(V) * \prod_{x \in X} obs(x).$$

This equation suggests the following effective way of entering observations:

Algorithm 5.5 (EnterEvidence) *Given observations $X = x$, for each $x \in X$, find a cluster Q in $\mathcal{T}$ such that $x \in Q$ and replace $B_Q(Q)$ by $B_Q(Q) * obs(x)$.*

After **EnterEvidence**, is performed the system potential will be

$$B_T(V) * \prod_{x \in X} obs(x).$$

If **UnifyBelief** is performed next, according to Theorem 5.16, $P(Q|x)$ can be obtained from each cluster Q by normalizing its potential $B_Q(Q)$.

To summarize, given a BN, it can be converted into a JT representation by moralization, triangulation, JT construction, and cluster potential assignment. Before any observation is available, **UnifyBelief** can be performed in the JT. The prior probability distribution of any variable v can then be obtained by marginalization of the potential of any cluster that contains v. After any observation $X = x$ is made, it can be entered by **EnterEvidence**. By performing **UnifyBelief** once again, the posterior probability distribution of any variable v can be obtained from any containing cluster. Our goal of exact belief updating for nontree BNs by concise message passing in a single graphical structure has now been achieved.

5.7 Bibliographical Notes

A special case of Theorem 5.8 is presented in Jensen [29] (Proposition 4.1). Another special case of Theorem 5.8 is presented in Shafer [62] (Section 1.2, Property 2). **Absorption**, **CollectEvidence**, and **DistributeEvidence** were proposed by Jensen et al. [30].

Alternative methods for exact belief updating with nontree BNs by concise message passing in a single graphical structure include Shafer–Shenoy propagation (Shafer [62]), laze propagation (Madsen and Jensen [39]), and propagation of belief functions by Shafer et al. [64].

5.8 Exercises

1. Calculate the product of the following two potentials.

Two potentials for Exercise 1

(x, y, z)	$B_1()$	(x, y, z)	$B_1()$	(x, w, z)	$B_2()$	(x, w, z)	$B_2()$
$(0, 0, 0)$	0.08	$(1, 0, 0)$	1.5	$(0, 0, 0)$	5.2	$(1, 0, 0)$	6.2
$(0, 0, 1)$	0.16	$(1, 0, 1)$	2.6	$(0, 0, 1)$	4.8	$(1, 0, 1)$	1.2
$(0, 1, 0)$	2.4	$(1, 1, 0)$	0	$(0, 1, 0)$	0.4	$(1, 1, 0)$	0
$(0, 1, 1)$	0.12	$(1, 1, 1)$	0.5	$(0, 1, 1)$	3.6	$(1, 1, 1)$	2.5

2. For the potentials $B_1(x, y, z)$ and $B_2(x, w, z)$ given in question 1, determine if $B_1(x, y, z)$ is zero-consistent with $B_2(x, w, z)$.
3. Compute the quotient $B_3(x, y, z)/B_4(x, w, z)$ with the potentials $B_3(x, y, z)$ and $B_4(x, w, z)$ given below:

Two potentials for Exercise 3

(x, y, z)	$B_3()$	(x, y, z)	$B_3()$	(x, w, z)	$B_4()$	(x, w, z)	$B_4()$
$(0, 0, 0)$	0.08	$(1, 0, 0)$	0	$(0, 0, 0)$	5.2	$(1, 0, 0)$	0
$(0, 0, 1)$	0.16	$(1, 0, 1)$	2.6	$(0, 0, 1)$	4.8	$(1, 0, 1)$	1.2
$(0, 1, 0)$	2.4	$(1, 1, 0)$	0	$(0, 1, 0)$	0.4	$(1, 1, 0)$	3.6
$(0, 1, 1)$	0.12	$(1, 1, 1)$	0.5	$(0, 1, 1)$	0.2	$(1, 1, 1)$	2.5

4. For the potentials given in question 3, compute

$$\sum_w B_3(x, y, z) * B_4(x, w, z).$$

(Hint: What is the most efficient method of computation?)

5. The topology of a digital circuit is shown in Figure 5.7. Each AND gate has a 0.01 failure rate. The failure rate for an OR gate is 0.015 and the one for a NOT gate is 0.005. When an AND gate fails, it produces correct output 10% of the time. When an OR gate fails, it produces correct output 40% of the time. When a NOT gate fails, it outputs correctly 70% of the time.

Construct a BN for the circuit. Use its JT representation to compute $P(v|a = 1, b = 0, h = 0)$ for each variable v that is unobserved.

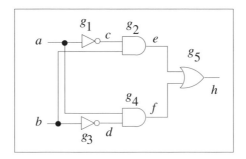

Figure 5.7: A digital circuit.

6. Proposition 5.13 allows the potential in C to change between the two absorptions but requires that the separator S remain supportive.

Under what situations, can a change in the potential of C render the originally supportive S unsupportive? Demonstrate with an example.

6

Multiply Sectioned Bayesian Networks

Chapters 2 through 5 studied exact probabilistic reasoning using a junction tree representation converted from a Bayesian network. The single-agent paradigm is followed in the study. Under this paradigm, a single computational entity, an agent, has access to a BN over a problem domain, converts the BN into a JT, acquires observations from the domain, reasons about the state of the domain by concise message passing over the JT, and takes actions accordingly. Such a paradigm has its limitations: A problem domain may be too complex, and thus building a single agent capable of being in charge of the reasoning task for the entire domain becomes too difficult. Examples of complex domains include designing intricate machines such as an aircraft and monitoring and troubleshooting complicated mechanisms such as chemical processes. The problem domain may spread over a large geographical area, and thus transmitting observations from many regions to a central location for processing is undesirable owing to communications cost, delay, and unreliability.

This and subsequent chapters consider the uncertain reasoning task under the *multiagent* paradigm in which a set of cooperating computational agents takes charge of the reasoning task of a large and complex uncertain problem domain. This chapter deals with the knowledge representation. A set of five basic assumptions is introduced to describe some ideal knowledge representation formalisms for multiagent uncertain reasoning. These assumptions are shown to give rise to a particular knowledge representation formalism termed *multiply sectioned Bayesian networks* (MSBNs). This places MSBNs at a prominent position in implementing multiagent distributed uncertain reasoning systems.

Section 6.2 defines the task of multiagent distributed uncertain reasoning. Section 6.3 introduces basic assumptions on agent belief representation, interagent message content, and agent organization preference. The logical implications of these assumptions for agent communications structure is derived by applying several results in earlier chapters to the multiagent context. Section 6.4 introduces a basic assumption about causal modeling and derives the implication for agent

107

communications interface. Section 6.5 combines and extends the results from Sections 6.3 and 6.4 to derive the dependence structure of a multiagent system. Section 6.6 introduces a basic assumption about agent knowledge complement and consistence. Multiply sectioned Bayesian networks are shown to be the logical consequence of the five basic assumptions. How to construct an MSBN, how to compile it for effective multiagent belief updating, and how to perform such belief updating by concise message passing will be presented in subsequent chapters.

6.1 Guide to Chapter 6

In this chapter we consider large and complex uncertain problem domains populated by multiple autonomous agents. The task of agents is to determine what is the true state of the domain so they can act upon it. To illustrate the ideas without getting into complex details, consider a *trivial* digital system (Figure 6.1) of three components: C_0, C_1, and C_2. Note that C_0 and C_1 are interfaced by signals j and k and that C_1 and C_2 are interfaced by signals a, b, and c. Suppose that three agents A_0, A_1, and A_2 are each assigned to one component. Their task is to monitor and troubleshoot the digital system. Each agent's knowledge is limited to one particular component (e.g., A_0 knows only the gates and their input and output relations in C_0 but knows nothing about the internal workings of C_1 beyond its sharing the signals j and k; A_0 does not even know the existence of C_2). Each agent's perspective is also limited so that it can only observe local events at the corresponding component. For instance, A_0 can observe the input and output of g_{10} with a cost, but it cannot observe the input and output of g_2 (it does not even know the existence of g_2). The state of each gate is unobservable. The input or output of some gates may also be unobservable. To determine the states of gates and unobservable signals, each agent must make inferences from its local observations. Because connections between components impose constraints on their signals, an agent may benefit from the knowledge and observations of other agents. For example, if j is unobservable and A_0 observes $k = 0$ and $l = 1$, then what A_1 thinks about the value of j (perhaps

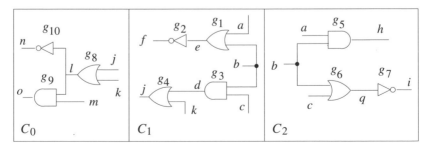

Figure 6.1: A trivial digital electronic system of three components.

from observations on d and k, and A_1's belief on the state of g_4) would be helpful for A_0 to determine the state of g_8. Section 6.2 elaborates on the task of distributed uncertain reasoning by multiple agents. In the remaining part of the chapter, we study how agents reasoning in such an uncertain environment should be organized in their knowledge representation and communication.

The approach taken is not to offer knowledge representation formalism as a solution that is to be taken for granted. Any such formalism must impose some constraints so that a given representation can be verified as legal or rejected as illegal. This is no different than the syntax of a programming language, which can be used to accept a given program as syntactically correct or to reject it as containing syntactic errors. A constraint, by definition, will restrict freedom in knowledge representation and hence may be considered undesirable for some reasons. Hence, the necessity of a constraint or a formalism cannot be determined by itself but needs to be judged by what it can deliver (i.e., what inference can be performed once the constraint is observed or the formalism is followed). The approach we take is then to lay out a set of seemingly desirable requirements (termed *basic assumptions*) and to derive a knowledge representation formalism as the logical consequence of these requirements.

As in the single-agent case, we represent the environment by a set of variables called a *total universe*. Because each agent only has a limited perspective on the environment, this implies that each agent only has knowledge over a subset of variables called a *subdomain*. For the preceding example, this means that A_0 only has the knowledge over $V_0 = \{j, k, l, m, n, o, g_8, g_9, g_{10}\}$.

Section 6.3 introduces three basic assumptions. The first of these stipulates that each agent's belief is represented by probability. This assumption not only requires each agent to represent its belief using a probability distribution but also to perform belief updating *exactly*. For the example above, this means that the belief of A_0 over the state of C_0 should be represented as $P_0(V_0)$. After $d = 0$ is observed by A_1 and communication between A_0 and A_1 takes place, A_0's belief should be $P_0(V_0|d = 0)$.

The second basic assumption requires an agent to communicate directly with another agent only with a *concise* message: its belief over the variables they share. Hence, A_0 can communicate directly with A_1 only with $P_0(j, k)$, and A_0 cannot communicate directly with A_2 because the two share no common variables. This basic assumption is consistent with the autonomy of agents. An agent needs to know nothing beyond its subdomain. Because each agent knows nothing beyond its subdomain, the only thing that can be exchanged is the belief on the common variables.

In Chapter 3, we have shown that exact belief updating cannot be achieved with concise message passing in cluster graphs with nondegenerate cycles. If we construct a cluster graph in which each cluster is the subdomain of an agent and

Figure 6.2: A cluster graph modeling the three-agent system.

each link is a nonempty intersection of two clusters it connects, then this cluster graph models the communication pathways of a multiagent system. For instance, the cluster graph in Figure 6.2 models the communication pathways of the three-agent system. The link $\langle V_0, V_1 \rangle$ signifies the direct communication path between A_0 and A_1, and the path $\langle V_0, V_1, V_2 \rangle$ signifies the indirect communication path between A_0 and A_2. Because cluster graphs model agent communication precisely, our analysis in Chapter 3 is directly applicable to multiagent communication. The third basic assumption requires that a simpler agent organization is preferred in which agent communication by concise message passing is achievable. According to the result from Chapter 3, it follows that a tree-organization for agent communication should be adopted.

Section 6.4 introduces the fourth basic assumption. It requires each agent to represent its subdomain dependence as a DAG. As demonstrated in Chapter 2, a DAG allows the agent's belief over a subdomain to be encoded concisely (through the chain rule). Hence, this assumption is a requirement about efficiency.

The first basic assumption requires each agent to represent its belief over its subdomain with a probability distribution. In theory, a JPD can also be defined over the total universe. What should be the relation between the JPD and each agent's belief over its subdomain? Section 6.6 introduces the fifth basic assumption, which requires the JPD to be consistent with each agent's belief over its subdomain. The assumption enforces cooperation among agents and interprets the JPD thus defined as the *collective belief* of all agents. In other words, the collective belief reflects the expertise of each agent within its subdomain and supplements the agent's limited local knowledge outside its subdomain.

From the five basic assumptions, Sections 6.3 through 6.6 derive stepwise the constraints that logically follow from them. Collectively, these constraints specify a knowledge representation formalism called multiply sectioned Bayesian networks (MSBNs). We illustrate the MSBN constructed for the three-agent example of Figure 6.2. Like a BN, an MSBN has its structural part. Each agent represents the dependence structure in its subdomain as a local DAG (G_0, G_1, and G_2 in Figure 6.3). The union of these local DAGs should be a connected DAG, as is the case in Figure 6.3 (although the union operation will never be performed in practice). The local DAGs should be organized into a tree such as the one in Figure 6.2 with each V_i replaced by G_i. This tree organization dictates the pathways of agent communication. A node shared by multiple local DAGs may have parent nodes in

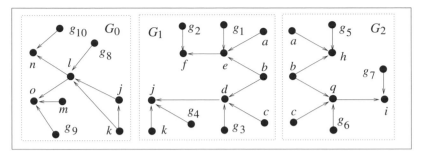

Figure 6.3: Local DAGs of the three-agent system.

several of them. All these parent nodes should be contained in at least one of the local DAGs. For instance, node j has parent k in G_0 and parents k, d, and g_4 in G_1. All of them are contained in G_1. The numerical part of the MSBN is very similar to that of a BN. To quantify the strength of probabilistic dependence, each node in the graphical structure is associated with a probability distribution conditioned on its parent variables. The exception is that, for each node shared by multiple agents, only one of its occurrences is associated with a conditional probability distribution. Figure 6.4 shows the conditional probability distributions associated with the local DAGs. Note that only one of the two occurrences of node j is assigned a conditional

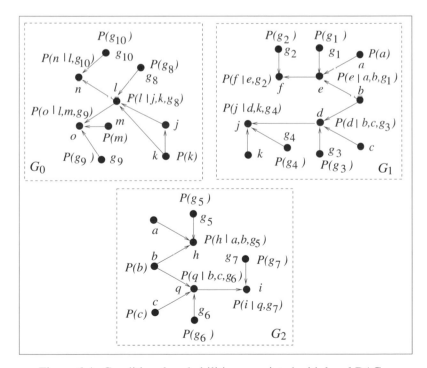

Figure 6.4: Conditional probabilities associated with local DAGs.

probability distribution. The collective belief of the multiagent system, the JPD, is then defined by the product of all such distributions.

6.2 The Task of Distributed Uncertain Reasoning

So far, we have considered modeling and inference using Bayesian networks under the single-agent paradigm. That is, a single reasoner holds a centralized knowledge representation, gathers observations about the domain, and reasons about the domain through a centralized inference computation. As discussed in Section 1.4, the larger and more complex the problem domain considered becomes the more difficult it is to use the single-agent paradigm. In the context of BNs, these problems include difficulty in constructing and maintaining a very large network as a single unit, the cost of transmitting observed variable values to a central location and the associated time delay and unreliability (due to communication failures), and the computational expense of belief updating in the large network. One solution is to explore the natural distribution of knowledge and sensors, and to distribute modeling and inference accordingly using the multiagent paradigm.

First, let us rephrase the uncertain reasoning task under the multiagent paradigm. Consider a large uncertain domain populated by multiple autonomous agents. The agents' task is to determine the true state of the domain so they can act upon it. We can still describe the domain with a set of variables, as in the single-agent paradigm. However, owing to the knowledge distribution and sensor distribution, each agent only has knowledge about a subset of variables, can only observe and reason about variables within the subset, and can only act to control variables within the subset. Clearly, to benefit from the knowledge and observations of others, agents must communicate. How can agents cooperate to accomplish the task with the least amount of communication?

To make the task concrete, consider a system of five components that may be remotely located. As shown in Figure 6.5, each component is interfaced with one or more additional components. The system may be mechanical, electrical, or chemical. It may be natural or man-made. We consider here a digital electronic system because no special domain knowledge from readers is required to comprehend the details. Figures 6.6 through 6.10 depict each of the five components C_0 through

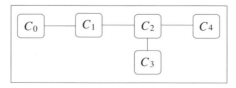

Figure 6.5: A system of five components.

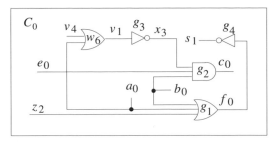

Figure 6.6: Component C_0 of a digital system.

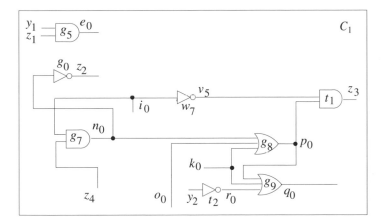

Figure 6.7: Component C_1.

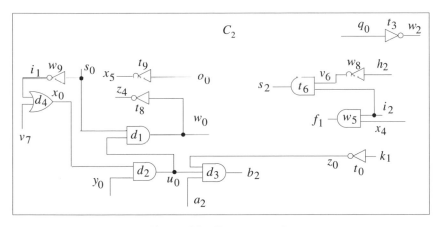

Figure 6.8: Component C_2.

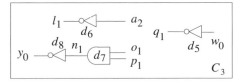

Figure 6.9: Component C_3.

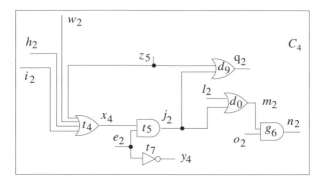

Figure 6.10: Component C_4.

C_4, and Figure 6.11 presents an integrated view of the entire system in which components are shown in dashed boxes.

Our purpose is to use this digital system to illustrate many issues arising in uncertain reasoning in multiagent systems. By making all devices in the system explicit, readers can better appreciate the technical issues to be presented and can verify the results manually. This will not be possible if, instead, a huge system is used with many details omitted. On the other hand, in practice, a digital system on the scale of the preceding example can easily be handled by a centralized monitoring agent. For instance, Bayesian networks with hundreds of variables have been built (e.g., Pradhan et al. [56] and Pfeffer et al. [53]). A multiagent system should be capable of handling problem domains of even higher complexity. Therefore, the foregoing example is not a reflection of the scale of the problems that we set out to deal with. Readers should keep this in mind because many of the examples used in the remaining chapters of the book apply the same compromise.

Each component can be associated with a computational agent responsible for monitoring and troubleshooting a given component. The agent can acquire local observations from sensors and reason about the values of unobservable variables in the component. Because components are interfaced with each other, observations obtained by other agents may be valuable to a given agent in reasoning about its local component. For instance, consider the part of the digital system depicted in Figure 6.12, which involves three components C_1, C_2, and C_3. Let the components be monitored by agents A_1, A_2, and A_3, respectively. Normally, the state of each gate is unobservable. That is, whether or not a gate is normal or faulty cannot be directly observed and can only be inferred from its input and output. Suppose that the output w_0 of the AND gate d_1 and the output z_4 of the NOT gate t_8 are also unobservable (or cannot be observed under an affordable cost). Now A_2 is unable to determine whether the gate d_1 is functioning correctly after the inputs s_0 and u_0 are observed. However, suppose that agent A_1 can observe i_0 and n_0 and that agent

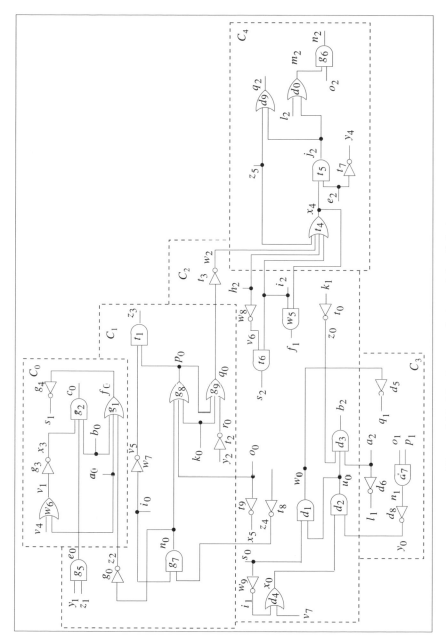

Figure 6.11: The integrated view of the digital system.

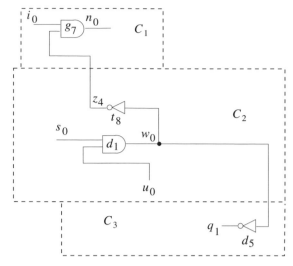

Figure 6.12: The agent for C_2 may benefit from the observations of other agents when w_0 and z_4 are not observable.

A_3 can observe q_1. Then, by communicating with A_1 and A_3, agent A_2 will be able to reduce its uncertainty about the gate d_1 and can sometimes reach a high level of certainty.

As a concrete scenario, suppose that A_2 observes $s_0 = 0$ and $u_0 = 1$, where 0 and 1 are logical signal levels (see Figure 6.13). Based on the knowledge about

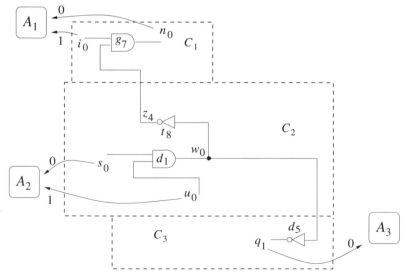

Figure 6.13: Agents A_1, A_2, and A_3 make observations at the corresponding components. Each arrow goes from an observable variable to an observing agent. The observed logical value is labeled beside the arrow head.

AND gates, A_2 expects w_0 to be 0 and z_4 to be 1 if gates d_1 and t_8 are normal. Agent A_2 cannot confirm the state of d_1 and t_8 because w_0 and z_4 are unobservable. Suppose that A_1 observes $i_0 = 1$ and $n_0 = 0$. If gate g_7 is normal, A_1 will expect z_4 (also unobservable for A_1) to be 0: a belief in conflict with that of A_2. Suppose that A_3 observes $q_1 = 0$. If gate d_5 is normal, A_3 will expect w_0 (also unobservable for A_3) to be 1, which is another belief conflicting with that of A_2. However, without communication among the three agents, the conflicts will not be realized.

Now suppose that A_2 communicates with the other two agents about their individual belief on shared variables: z_4 between A_1 and A_2 and w_0 between A_2 and A_3. The conflicts will now be perceived. Clearly, the observations imply that it is impossible for all gates to be functioning normally, as originally believed. Several hypotheses, however, exist. One hypothesis is that d_1 is normal and produces 0, z_4 is normal and produces 1, but gates g_7 and d_5 are both faulty. Another hypothesis is that d_1 is faulty and produces 1 and all other gates are normal. If gates break down independently with small probabilities, as is usually the case, the second hypothesis (one faulty gate) is much more likely than the first one (two faulty gates). Therefore, unless there are observations in the rest of the digital system that provide additional support to the first hypothesis, the agents should believe the second hypothesis much more than the first. On the basis of that belief, A_2 perhaps would either sound an alarm about d_1 (which requires little additional hardware resources), or activate a backup circuit (which requires significantly more additional resources) to replace d_1.

Our objective is to develop a computational framework that will allow such multiagent uncertain reasoning to be performed. In the following sections, we specify several basic assumptions about the multiagent reasoning task and derive a representation formalism, MSBNs, as the logical consequence of these assumptions. The term *communication* will be used to refer to any exchange of messages, directly or indirectly, between two or more agents.

6.3 Organization of Agents during Communication

Following Bayesian probability theory, we assume that each agent represents its belief about the problem domain in terms of a probability distribution.

Basic Assumption 6.1 *Each agent's belief is represented by probability.*

We also interpret Basic Assumption 6.1 as requiring belief updating to be performed exactly (versus approximately).

We consider a total universe V of variables over which a multiagent system made of $n > 1$ agents $A_0, \ldots, A_{n-1}$ is defined. Each A_i has knowledge over a $V_i \subset V$, called the *subdomain* of A_i. Note that $\cup_i V_i = V$. From Basic Assumption 6.1, the knowledge of A_i is a probability distribution over V_i denoted by $P_i(V_i)$.

We take it for granted that for agents A_i and A_j to communicate *directly* they must share some variables, that is, $V_i \cap V_j \neq \emptyset$. We will refer to $V_i \cap V_j$ as the *interface* between agents A_i and A_j. To minimize communication, we allow agents to exchange only their belief on the interface. Although an agent can communicate directly only through an interface, it can influence another agent by *indirect* communication through a sequence of direct communications over a set of intermediate agents. In a multiagent system, each agent's belief should potentially be influential in any other, directly or indirectly; otherwise, the system can be split in two. We express these considerations in the following basic assumption.

Basic Assumption 6.2 *An agent A_i can in general influence the belief of each other agent through direct or indirect communication but can communicate* directly *to another agent A_j only with $P_i(V_i \cap V_j)$ where $V_i \cap V_j \neq \emptyset$.*

Basic Assumption 6.2 requires *concise* (versus *massive*) messages, $P_i(V_i \cap V_j)$, to be passed between agents. Note that it is implied that A_i can also communicate directly with A_j by receiving $P_j(V_i \cap V_j)$ from the latter. To distinguish this communication from the message sent between nodes or clusters in earlier chapters, we refer to it as an *internal message* or simply *i-message*, and we refer to $P_i(N_i \cap N_j)$ as an *external message* or simply *e-message*. Accordingly, we term direct communication between agents *e-message passing*. Note that both i-messages and e-messages are *concise* messages (versus *massive* messages in Section 2.8).

Consider the digital system example. According to the physical division of components, the subdomain for agent A_0 is

$$V_0 = \{a_0, b_0, c_0, e_0, f_0, g_1, g_2, g_3, g_4, s_1, v_1, v_4, w_6, x_3, z_2\},$$

and the subdomain for agent A_1 is

$$V_1 = \{e_0, g_0, g_5, g_7, g_8, g_9, i_0, k_0, n_0, o_0, p_0, q_0, r_0, t_1, t_2, v_5, w_7,$$
$$y_1, y_2, z_1, z_2, z_3, z_4\}.$$

The two subdomains share variables

$$V_0 \cap V_1 = \{e_0, z_2\},$$

because they are the output of component C_1 into component C_0. Such subdomain division, entirely determined by the physical component division, although sometimes preferred, is less general and inflexible.

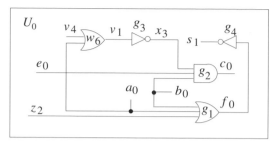

Figure 6.14: Virtual component U_0 of the digital system.

Subdomain division does not have to be bounded by the actual physical division. Some physical devices (gates in the example) in one component can be "virtually" represented in subdomains associated with other components. This more general representation is sometimes preferred because it is desirable to share certain variables among agents beyond the physical input or output. It may also help to achieve certain technical properties, as will be seen in the later chapters (e.g., Section 10.6). We will demonstrate this more general subdomain division with the understanding that one can always apply the more restrictive physical division if such generality is not needed. Figures 6.14 through 6.17 depict the "virtual" components U_0 through U_3.

The virtual component U_0 is identical to the physical component C_0. For the convenience of comparison with U_1, we reproduce U_0 here. The virtual component U_1, however, differs from its physical counterpart C_1. The "virtual" devices are indicated by a dashed box. Similarly, virtual components U_2 and U_3 also contains

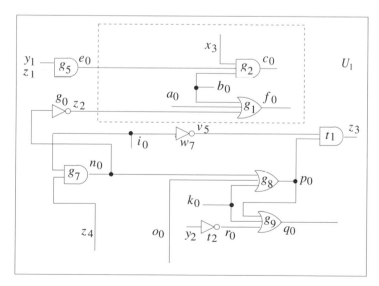

Figure 6.15: Virtual component U_1 of the digital system.

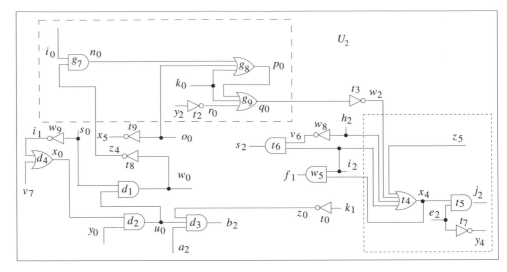

Figure 6.16: Virtual component U_2.

virtual devices. The virtual component U_4 is identical to the physical component C_4 and hence is not shown. Note that an agent cannot observe a virtual signal. If a signal is represented in more than one component, only its physical occurrence may be observable. For example, the signal u_0 is physically present in component C_2 but is represented in both virtual components U_2 and U_3. Only the agent A_2 (but not A_3) may observe the value of u_0 if the signal is observable. The subdomain division using the virtual components is summarized in Table 6.1. The resultant agent interfaces are summarized in Table 6.2.

Pathways for e-message passing can be represented by a junction graph (Section 3.2) $H = (V, \Omega, E)$, where $\Omega = \{V_i | i = 0, \ldots, n - 1\}$, and

$$E = \{\langle V_1, V_2 \rangle | V_1, V_2 \in \Omega, V_1 \neq V_2, V_1 \cap V_2 \neq \emptyset\}.$$

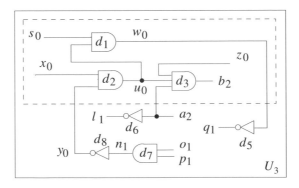

Figure 6.17: Virtual component U_3.

Table 6.1: *Subdomains of digital system monitoring agents*

Agent	Component	Subdomain
A_0	U_0	$V_0 = \{a_0, b_0, c_0, e_0, f_0, g_1, g_2, g_3, g_4, s_1, v_1, v_4, w_6, x_3, z_2\}$
A_1	U_1	$V_1 = \{a_0, b_0, c_0, e_0, f_0, g_0, g_1, g_2, g_5, g_7, g_8, g_9, i_0, k_0, n_0,$ $o_0, p_0, q_0, r_0, t_1, t_2, v_5, w_7, x_3, y_1, y_2, z_1, z_2, z_3, z_4\}$
A_2	U_2	$V_2 = \{a_2, b_2, d_1, d_2, d_3, d_4, e_2, f_1, g_7, g_8, g_9, h_2, i_0, i_1, i_2,$ $j_2, k_0, k_1, n_0, o_0, p_0, q_0, r_0, s_0, s_2, t_0, t_2, t_3, t_4, t_5,$ $t_6, t_7, t_8, t_9, u_0, v_6, v_7, w_0, w_2, w_5, w_8, w_9, x_0, x_4, x_5,$ $y_0, y_2, y_4, z_0, z_4, z_5\}$
A_3	U_3	$V_3 = \{a_2, b_2, d_1, d_2, d_3, d_5, d_6, d_7, d_8, l_1, n_1, o_1, p_1, q_1, s_0,$ $u_0, w_0, x_0, y_0, z_0\}$
A_4	U_4	$V_4 = \{d_0, d_9, e_2, g_6, h_2, i_2, j_2, l_2, m_2, n_2, o_2, q_2, t_4, t_5, t_7,$ $w_2, x_4, y_4, z_5\}$

Each separator in H represents a pathway for direct interagent communication. Each path of length greater than 1 represents a pathway for indirect communication. We will refer to H as a *communications graph*. Figure 6.18 shows the communications graph for the digital system monitoring agents.

Although this graph turns out to be a tree, it is not the case in general. For instance, the virtual component U_2 may alternatively be defined as U_2' shown in Figure 6.19. The additional virtual device and signal are shown in the shaded area. This addition not only enlarges the interface between A_1 and A_2 into

$$I_{1,2}' = \{\underline{g_0}, g_7, g_8, g_9, i_0, k_0, n_0, o_0, p_0, q_0, r_0, t_2, y_2, \underline{z_2}, z_4\},$$

where the new elements are underlined, but also creates a new interface between A_0 and A_2,

$$I_{0,2}' = \{z_2\}.$$

The communications graph, after U_2 is replaced by U_2', is shown in Figure 6.20.

Table 6.2: *Agent communication interfaces*

Agents	Interface
A_0 vs A_1	$I_{0,1} = \{a_0, b_0, c_0, e_0, f_0, g_1, g_2, x_3, z_2\}$
A_1 vs A_2	$I_{1,2} = \{g_7, g_8, g_9, i_0, k_0, n_0, o_0, p_0, q_0, r_0, t_2, y_2, z_4\}$
A_2 vs A_3	$I_{2,3} = \{a_2, b_2, d_1, d_2, d_3, s_0, u_0, w_0, x_0, y_0, z_0\}$
A_2 vs A_4	$I_{2,4} = \{e_2, h_2, i_2, j_2, t_4, t_5, t_7, w_2, x_4, y_4, z_5\}$

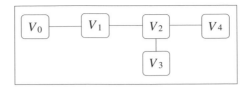

Figure 6.18: The communications graph for digital system monitoring.

Because each agent's belief can potentially be influential in any other agent's belief, directly or indirectly, *H* is *connected*. This analysis is summarized in Proposition 6.1.

Proposition 6.1 *Let H be the communications graph of a multiagent system that observes Basic Assumptions 6.1 and 6.2. Then H is connected.*

The communications graph of a multiagent system is graph-theoretically equivalent to the cluster graph we investigated in Chapter 3 for single-agent i-message passing, as summarized in Table 6.3. This equivalence is important, for it makes several results on cluster graphs in earlier chapters directly applicable to communications graphs, as we shall see later. On the other hand, communications graphs and cluster graphs are semantically different. A subdomain in a communications graph is at the abstraction level of an agent, but a cluster in a cluster graph is at the subagent abstraction level. This difference leads to several new opportunities

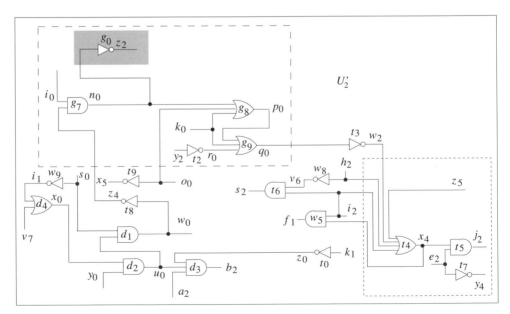

Figure 6.19: An alternative virtual component U_2'.

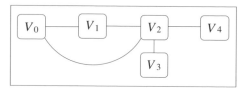

Figure 6.20: The communications graph for digital system monitoring after the virtual component U_2 is redefined as U_2'.

as well as issues that are unique for multiagent systems, which will be explored in this and subsequent chapters.

First, the graph-theoretical equivalence allows us to apply Corollary 3.5 directly to the multiagent communication and arrive at the following conclusion: In general, belief updating of multiple agents cannot be achieved in a communications graph with nondegenerate cycles, no matter how e-message passing is performed. Although communications graphs with strong degenerate cycles and some with weak ones admit belief updating by e-message passing, these communications graphs can be substituted by cluster trees. Section 3.6 shows that substitution of cluster trees for cluster graphs with degenerate cycles do not affect belief updating. This implies that multiagent belief updating can be achieved by organizing e-message passing in a cluster tree of the communications graph. Because the cluster tree is the simplest subgraph retaining connectedness, computation is more efficient. Hence, we prefer a simpler organization of agents when degenerate cycles exist in the communications graph.

Basic Assumption 6.3 *A simpler agent organization (as a subgraph of the communications graph) is preferred.*

From the equivalence of communications graph and cluster graph and the analyses in Sections 3.5 and 3.6, it follows that agents and their subdomains should be organized into a *tree*, as summarized in Proposition 6.2.

Proposition 6.2 *Let a multiagent system be one that observes Basic Assumptions 6.1 through 6.3. Then a tree organization of agents must be used for communicating of beliefs.*

Table 6.3: *Correspondence between a cluster graph and a communications graph*

Cluster graph	Communications graph
domain	total universe
cluster	subdomain
separator	interface

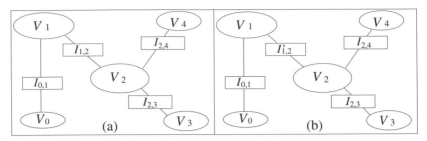

Figure 6.21: Junction tree organizations for agent communication.

Theorem 3.7 in Section 3.7 has established that local consistency in a cluster tree ensures global consistency if and only if the cluster tree is a junction tree. The i-message passing in a cluster tree can be used to achieve local consistency between adjacent clusters (over their separator belief). Similarly, e-message passing can be used to achieve consistency between adjacent agents (over their interface belief) in their tree organization. Given the equivalence of communications graph and cluster graphs, the result from Theorem 3.7 implies that consistency between adjacent agents ensures consistency among all agents if and only if the agents are organized into a junction tree. This is summarized in Proposition 6.3.

Proposition 6.3 *Let a multiagent system be one that observes Basic Assumptions 6.1 through 6.3. Then a junction tree organization of agents must be used for communicating beliefs.*

The junction tree organization of agents for monitoring the digital system is shown in Figure 6.21 with subdomains and interfaces specified in Tables 6.1 and 6.2. The junction tree in (a) corresponds to virtual components U_0 through U_4, and the one in (b) corresponds to virtual components with U_2 replaced by U_2'. Note that (a) is isomorphic to the communications graph in Figure 6.18 because it is itself a junction tree. On the other hand, (b) is *not* isomorphic to the communications graph in Figure 6.20. Note that the cycle in Figure 6.20 is a weak degenerate cycle and must be broken at the separator between V_0 and V_2, which produces Figure 6.21(b). When agents are organized into a junction tree, we refer to a pair of agents as being *adjacent* according to their topological locations on the junction tree. For instance, agents A_2 and A_4 are adjacent in the organization of Figure 6.21.

6.4 Agent Interface

Each link in a tree splits the tree into two subtrees. Given our commitment to a junction tree agent organization (Proposition 6.3), it follows that the interface between each pair of adjacent agents forms the unique communications channel between the two groups of agents each located in one subtree. Let Z denote any interface, X represent the union of all subdomains of one subtree induced by the interface excluding Z, and Y signify the union of all subdomains of the other subtree excluding Z. By Basic Assumption 6.2, $P(Z)$ is the only information that can be directly communicated between X and Y. Note that because we are concerned with Z as the interface between X and Y, we can safely ignore the distribution of X (or Y) among multiple agents. This abstraction makes Proposition 4.5 immediately applicable. It implies that, with other conditions satisfied (namely, with $P(X, Y, Z)$, $P(X, Z)$, and $P(Y, Z)$ properly defined, which will be discussed in Section 6.6), in order to perform belief updating by e-message passing over interface, the interface Z must render X and Y conditionally independent, that is, $I(X, Z, Y)$ must hold. This is summarized in the following proposition.

Proposition 6.4 *Let a multiagent system be one that observes Basic Assumptions 6.1 through 6.3. Then each interface in the tree organization must render subdomains in the two induced subtrees conditionally independent.*

Proposition 6.4 essentially imposes a *semantic* constraint on how to divide the total universe into subdomains among agents. Although satisfaction of the restraint cannot be guaranteed entirely through syntactic means, it does imply certain syntactic constraints, as will be seen.

In a complex domain, each subdomain can still be large, and hence the representation using a graphical model is justified. We assume that a DAG is used to structure the agent's knowledge about a subdomain. In other words, we are committed to causal modeling of the agent's knowledge.

Basic Assumption 6.4 *A DAG is used to structure each agent's knowledge.*

Note that this assumption is not essential. For example, a JT representation of the subdomain knowledge can also be used. In Chapter 7, the local DAG representation of each agent will be converted into a JT representation for effective inference in a way similar to the conversion studied in Chapter 4. Because knowledge acquisition can be performed more compactly using the DAG structure, we will use this approach here.

A DAG model admits a causal interpretation of dependence. Once we adopt it for each agent, we must adopt it for the joint representation of all agents.

Proposition 6.5 *Let a multiagent system be one that observes Basic Assumptions 6.1 through 6.4. Then each subdomain V_i is structured as a DAG over V_i, and the union of these DAGs is a connected DAG over V.*

Proof: If the union of subdomain DAGs is not a DAG, then it has a directed cycle. This contradicts the causal interpretation of individual DAG models. The connectedness is implied by Proposition 6.1.   □

To avoid confusion, we refer to a DAG defined over a subdomain as a *local DAG* and to the union of local DAGs as the *DAG union* of the multiagent system.

We are committed to conditionally independent agent interfaces and DAG subdomain modeling. As interface variables are embedded in subdomain DAGs, what topological constraint must they satisfy, if any? Recall from Section 4.4 that the graphical dependence structure should be an I-map of the problem domain. To encode the conditional independence induced by each agent interface, that is, $I(X, Z, Y)$, the interface variables should graphically separate X and Y in the DAG union, that is, $\langle X, Z, Y \rangle$. For directed graph structures, the criterion for graphical separation is d-separation (Section 4.2). It then follows that Z should d-separate X and Y. We define the concept of a *d-sepset* below and show that each agent interface should be a d-sepset.

Definition 6.6 *Let* $G_i = (V_i, E_i) (i = 0, 1)$ *be two DAGs such that* $G = G_0 \sqcup G_1$ *is a DAG. A node* $x \in I = V_0 \cap V_1$ *with its parents* $\pi(x)$ *in* G *is a **d-sepnode** between* G_0 *and* G_1 *if either* $\pi(x) \subseteq V_0$ *or* $\pi(x) \subseteq V_1$. *If every* $x \in I$ *is a d-sepnode, then* I *is a **d-sepset**.*

Figure 6.22 illustrates the d-sepset concept. The agent interface is $\{e, f\}$. For the shared variable f, we have $\pi(f) = \{c, e\}$ and $\pi(f) \subset V_0$. Hence, f is a d-sepnode. For e, we have $\pi(e) = \{a, d\}, \pi(e) \not\subset V_0$ and $\pi(e) \not\subset V_1$. Hence, e is not a d-sepnode and $\{e, f\}$ is *not* a d-sepset. If, however, the arc (a, e) in G_0 were reversed, then $\{e, f\}$ would be a d-sepset.

Note also that the condition

$$\text{either } \pi(x) \subseteq V_0 \quad \text{or} \quad \pi(x) \subseteq V_1$$

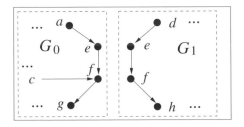

Figure 6.22: An agent interface $\{e, f\}$ that is not a d-sepset. Each box represents the DAG of one agent. Dots represent additional variables not shown explicitly.

in Definition 6.6 defines a *disjunction* and is true if

$$\pi(x) \subseteq V_0 \quad \text{and} \quad \pi(x) \subseteq V_1.$$

For instance, if the arc (c, f) in Figure 6.22 were reversed, then we would have $\pi(f) = \{e\}$, $\pi(f) \subset V_0$, and $\pi(f) \subset V_1$. The node f would still be a d-sepnode. Similarly, if the arcs (a, e) and (d, e) were both reversed, we would have $\pi(e) = \emptyset$, $\pi(e) \subset V_0$ and $\pi(e) \subset V_1$. The node e would become a d-sepnode.

The following proposition establishes that each agent interface Z should be a d-sepset.

Proposition 6.7 *Let $G_i = (V_i, E_i)$ ($i = 0, 1$) be two DAGs such that $G = G_0 \sqcup G_1$ is a DAG. Then $V_0 \backslash V_1$ and $V_1 \backslash V_0$ are d-separated by $I = V_0 \cap V_1$ if and only if I is a d-sepset.*

Proof: [Sufficiency] Suppose I is a d-sepset. Let ρ be a path between $V_0 \backslash V_1$ and $V_1 \backslash V_0$ such that all nodes in one side of ρ belong to $V_0 \backslash I$, all nodes on the other side belong to $V_1 \backslash I$, and one or more adjacent d-sepnodes in I are in between. It suffices to show that every such path is closed by I. Every ρ has at least one d-sepnode. By Definition 4.2, if one d-sepnode on ρ is tail-to-tail or head-to-tail, then ρ is closed by I.

Consider the case in which ρ has only one d-sepnode x. We show that x cannot be head-to-head on ρ. Suppose x is head-to-head with parents y and z on ρ. Because x is the only d-sepnode on ρ, neither y nor z is shared by V_0 and V_1, say, $y \in V_0$ and $z \in V_1$. This means $\{y, z\} \not\subseteq V_0$ and $\{y, z\} \not\subseteq V_1$, which implies that x is not a d-sepnode: a contradiction. Hence, x is either tail-to-tail or head-to-tail on ρ.

Next, consider the case in which ρ contains at least two d-sepnodes. We show that one of them cannot be head-to-head on ρ. Pick two d-sepnodes x and y on ρ that are adjacent. Such d-sepnodes do exist because of the way ρ is defined. The x and y are connected either by (x, y) or by (y, x). In either case, one of them must be a tail.

[Necessity] Suppose there exists $x \in I$ with distinct parents y and z in G such that $y \in V_0$ but $y \notin V_1$, and $z \in V_1$ but $z \notin V_0$. Note that the condition disqualifies I from being a d-sepset, and this is the only way that I may become disqualified. Now y and z are not d-separated by x and hence $V_0 \backslash V_1$ and $V_1 \backslash V_0$ are not d-separated by I. $\square$

Proposition 6.7 gives us a syntactic rule to detect if an agent interface does not conform to the conditional independence requirement. The d-sepset constraint disallows interface variables such as e in Figure 6.22. This can be understood as follows: When the value of e or any descendant of e (e.g., h) is observed, it induces dependence between a and d. Now knowing the value of d can change an agent's belief on a. That is, the information on the interface $\{e, f\}$ is no longer sufficient for one agent to inform the other.

In summary, the d-sepset constraint is necessary (and syntactically sufficient) to ensure correct e-message passing in all possible observation patterns. This constraint may be relaxed if certain observation patterns will never occur. We will leave this possibility for speculation.

6.5 Multiagent Dependence Structure

So far, we have made four basic assumptions about agent belief, interagent message passing, preference of simpler agent organization, and causal modeling. We have shown that several representational constraints follow logically from these assumptions, including connectivity of agent communications graph, junction tree agent organization, and d-sepset agent interface. These basic assumptions and logical consequences collectively dictate a multiagent dependence structure at two levels: the agent level modeling and the society–system level organization. We introduce the *hypertree* notion to describe the relation between the two levels.

Definition 6.8 *Let $G = (V, E)$ be a connected graph sectioned into subgraphs $\{G_i = (V_i, E_i)\}$. Let the G_is be organized as a connected tree Ψ, where each node is labeled by a G_i and each link between G_k and G_m is labeled by the interface $V_k \cap V_m$ such that for each i and j, $V_i \cap V_j$ is contained in each subgraph on the path between G_i and G_j in Ψ. Then Ψ is a **hypertree** over G. Each G_i is a **hypernode**, and each interface is a **hyperlink**.*

An example hypertree is shown in Figure 6.23.

A node in G may be shared by more than two subgraphs. To ensure d-separation by interfaces, we need to extend the d-sepset definition in Section 6.4, which is based only on a *pair* of subgraphs. Consider Figure 6.24 in which G is sectioned into $\{G_i | i = 0, \dots, 4\}$. Each hyperlink in Ψ satisfies Definition 6.6 for a d-sepset. However, let $X = \{q, r\}$, $Z = \{j, k, l\}$, and $Y = V \backslash \{j, k, l, q, r\}$; then, $\langle X, Z, Y \rangle_G$ does not hold according to d-separation, because the path $\langle p, j, q \rangle$ is rendered open by Z as well as the path $\langle i, j, q \rangle$. Such agent interfaces are problematic. For instance, the message that A_4 can send to A_3 is $P_4(j, k, l)$ by Basic Assumption 6.2. If j, k, l, and q are all observed by A_4, then the message is identical no matter what value of q is observed. However, because i and q are not d-separated by Z and hence $I(i, Z, q)$ does not hold, A_2's belief on i should in general be different when a different value of q is observed by A_4. The agent interfaces fail in this case.

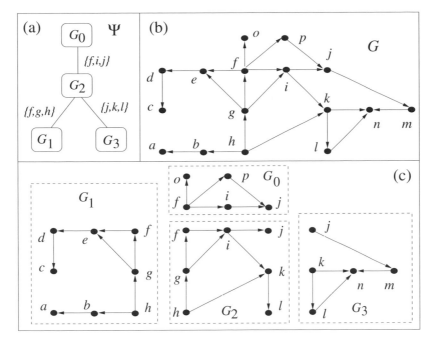

Figure 6.23: Here Ψ in (a) is a hypertree over G in (b), which is sectioned into subgraphs in (c).

To avoid such failure, we may extend Definition 4.2 in an obvious manner and require the agent interface to satisfy the following condition:

Condition 6.9 *Let G be a directed graph such that a hypertree over G exists. A node x contained in more than one subgraph in G must be such that for every pair of subgraphs G_i and G_j that contain x with the parents $\pi_{ij}(x)$ in $G_i \sqcup G_j$, either $\pi_{ij}(x) \subseteq V_i$ or $\pi_{ij}(x) \subseteq V_j$.*

Apply this condition to the example in Figure 6.24. Although G_0 and G_4 are not adjacent on the hypertree Ψ, they are now subject to the test of whether

$$\pi_{04}(j) \subseteq V_0 \quad \text{or} \quad \pi_{04}(j) \subseteq V_4$$

holds. Because $\pi_{04}(j) = \{i, p, q\}$, neither $\pi_{04}(j) \subseteq V_0$ nor $\pi_{04}(j) \subseteq V_4$ holds, and hence j is not a valid interface node.

Although problematic agent interfaces such as those in Figure 6.24 are invalidated by Condition 6.9, the condition is too strong and disallows some valid agent interfaces. Consider the hypertree DAG union in Figure 6.25. The node x is shared by all three subgraphs. Its parents in G_0 and G_1 are $\pi_{01}(x) = \{y, z\}$, and neither $\pi_{01}(x) \subseteq V_0$ nor $\pi_{01}(x) \subseteq V_1$ holds. That is, x does not satisfy Condition 6.9.

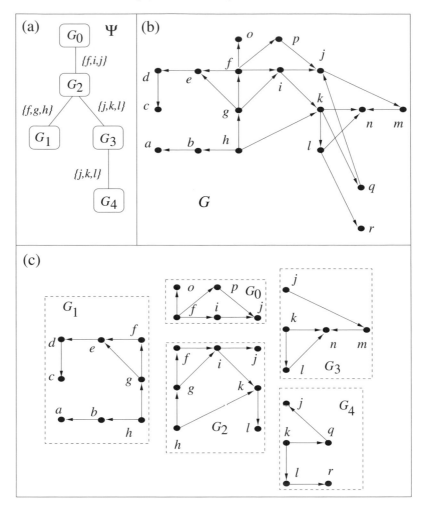

Figure 6.24: Here Ψ in (a) is a hypertree over G in (b), which is sectioned into subgraphs in (c).

On the other hand, the agent interface $I_{1,2} = \{x, z\}$ d-separates $V_1 \backslash I_{1,2}$ and $(V_0 \cup V_2) \backslash I_{1,2}$, and the agent interface $I_{0,2} = \{x, y\}$ d-separates $V_0 \backslash I_{0,2}$ and $(V_1 \cup V_2) \backslash I_{0,2}$. Because $G_1 - G_2 - G_0$ is the only possible hypertree given the graph sectioning, A_0 can only communicate indirectly with A_1 through A_2. Hence, the agent interfaces are valid with respect to Proposition 6.4 (if the graph dependence structure is an I-map).

To avoid invalidating agent interfaces such as those in Figure 6.25, Condition 6.9 needs to be weakened. The following Proposition 6.10 establishes a necessary condition of Condition 6.9, which turns out to define the d-sepset on a hypertree properly.

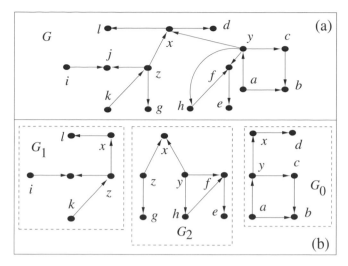

Figure 6.25: Graph G (a) is sectioned into subgraphs (b) with the hypertree $G_1 - G_2 - G_0$.

Proposition 6.10 *Let G be a directed graph satisfying Condition 6.9. Then for each node x contained in more than one subgraph, there exists a subgraph that contains all the parents $\pi(x)$ of x in G.*

Proof: We prove by contradiction. Suppose that Condition 6.9 holds in G, but for a given x, there exists no single subgraph containing $\pi(x)$. Then $y, z \in \pi(x)$ and two subgraphs G_i and G_j exist such that $y \in V_i$, $z \in V_j$, $y \notin V_j$, and $z \notin V_i$. Clearly, $y, z \in \pi_{ij}(x)$. Hence, we have found a pair of subgraphs G_i and G_j such that neither $\pi_{ij}(x) \subseteq V_i$ nor $\pi_{ij}(x) \subseteq V_j$, which contradicts Condition 6.9.   □

Using the weaker condition in Proposition 6.10, we extend the d-sepset in a hypertree as follows:

Definition 6.11 *Let G be a directed graph such that a hypertree over G exists. A node x contained in more than one subgraph with its parents $\pi(x)$ in G is a **d-sepnode** if there exists one subgraph that contains $\pi(x)$. An interface I is a **d-sepset** if every $x \in I$ is a d-sepnode.*

Definition 6.11 classifies the two representative hypertree DAG unions above correctly. According to the definition, not all agent interfaces in Figure 6.24 satisfy the d-sepset condition, for there is no subgraph containing all parents of node j. On the other hand, the agent interfaces in Figure 6.25 satisfy the d-sepset condition because G_2 contains all parents of x in G, G_0 contains all parents of y, and G_1 contains all parents of z. The following theorem establishes in general that Definition 6.11 is both necessary and sufficient. It is a generalization of Proposition 6.7 to a hypertree DAG union.

Theorem 6.12 *Let Ψ be a hypertree over a directed graph $G = (V, E)$. For each hyperlink I that splits Ψ into two subtrees over $U \subset V$ and $W \subset V$, respectively, $U \setminus I$ and $W \setminus I$ are d-separated by I if and only if each hyperlink in Ψ is a d-sepset.*

Proof: [Sufficiency] Assume that each hyperlink is a d-sepset. We show that for any given hyperlink I, $U \setminus I$ and $W \setminus I$ are d-separated by I.

Let ρ be a path between $U \setminus I$ and $W \setminus I$ such that all nodes on one side of ρ belong to $U \setminus I$, all nodes on the other side belong to $W \setminus I$, and one or more adjacent nodes in I are in between. It suffices to show that every such path is closed by I. Every ρ has at least one d-sepnode. By Definition 4.2, if one d-sepnode on ρ is tail-to-tail or head-to-tail, then ρ is closed by I.

Consider the case in which ρ has only one d-sepnode x. We show that x cannot be head-to-head on ρ. Suppose x is head-to-head with parents y and z on ρ. Because x is the only d-sepnode on ρ, neither y nor z is shared by U and W, say, $y \in U$ and $z \in W$. This means $\{y, z\} \not\subset U$ and $\{y, z\} \not\subset W$. Given that x is a d-sepnode, there exists a subgraph G_k that contains $\pi(x)$. Inasmuch as G_k is either located in the subtree over U or the subtree over W, either $\pi(x) \subset U$ or $\pi(x) \subset W$ holds. Because $\{y, z\} \subseteq \pi(x)$, it follows that either $\{y, z\} \subset U$ or $\{y, z\} \subset W$ must hold, which is a contradiction. Hence, x is either tail-to-tail or head-to-tail on ρ.

Next, consider the case in which ρ contains at least two d-sepnodes. We show that one of them cannot be head-to-head on ρ. Pick two d-sepnodes x and y on ρ that are adjacent. Such x and y do exist owing to the way ρ is defined. The x and y are connected either by (x, y) or by (y, x). In either case, one of them must be a tail node.

[Necessity] Assume that every hyperlink d-separates the two subtrees. We show that each hyperlink is a d-sepset by contradiction.

Suppose that there exists a shared node x such that no subgraph contains $\pi(x)$ (hence not every hyperlink is a d-sepset). Then there exists a hyperlink I on Ψ, where $x \in I$, and there are nonempty subsets

$$\pi_U(x) \subset \pi(x) \quad \text{and} \quad \pi_W(x) \subset \pi(x),$$

such that $\pi_U(x) \subset U$ and $\pi_W(x) \subset W$, $\pi_U(x)$ is incomparable with $\pi_W(x)$, and $\pi_U(x) \cup \pi_W(x) = \pi(x)$. Because $\pi_U(x)$ is incomparable with $\pi_W(x)$, there exist $y \in \pi_U(x)$ but $y \notin \pi_W(x)$ and $z \in \pi_W(x)$ but $z \notin \pi_U(x)$. The path $\rho = \langle y, x, z \rangle$ between U and W is rendered open by I because x is head-to-head on ρ. Hence, $U \setminus I$ and $W \setminus I$ are not d-separated by I, which is a contradiction.   □

Using the hypertree concept, we define the *hypertree multiply sectioned DAG* (or hypertree MSDAG). We show in the remainder of this section that it is the multiagent dependence structure logically implied by the basic assumptions.

Definition 6.13 *A **hypertree MSDAG** $G = \sqcup_i G_i$, where each $G_i = (V_i, E_i)$ is a DAG, is a connected DAG such that (1) there exists a hypertree Ψ over G, and (2) each hyperlink in Ψ is a d-sepset.*

Because the DAGs in Figure 6.23(c) form a hypertree and each interface is a d-sepset, they collectively form a hypertree MSDAG, whereas Figure 6.24(c) is not a hypertree MSDAG. The following proposition establishes the necessity of a hypertree MSDAG as the multiagent dependence structure:

Proposition 6.14 *Let a multiagent system be one that observes Basic Assumptions 6.1 through 6.4. Then it is structured as a hypertree MSDAG.*

Proof: The requirement for individual DAGs and the connected DAG union follows from Proposition 6.5. The hypertree organization follows from Proposition 6.3. The d-sepset condition follows from Proposition 6.4, Proposition 6.12, and our commitment to encode dependence relations graphically through I-maps. ☐

6.6 Multiply Sectioned Bayesian Networks

By Propositions 6.14, the dependence structure of a multiagent system is a connected DAG G. Hence, a JPD over V can be defined by specifying a local distribution for each node and applying the chain rule. Each node in G is either internal to an agent (non–d-sepnode) or is shared (d-sepnode).

The distribution for an internal node can be specified by the corresponding agent vendor. On the other hand, when a node is shared, its parent set within each agent may differ. In Figure 6.23(c), j has no parent in G_3, one parent in G_2, and two parents in G_0. Because each shared node x is a d-sepnode, by Definition 6.11 there exists an agent that contains all parents $\pi(x)$ of x in G.

When agents that contain x are developed by the same vendor, only $P(x|\pi(x))$ needs to be specified. For each agent A_k that contains $\pi_k(x) \subset \pi(x)$, $P_k(x|\pi_k(x))$ is implied (see Exercise 5) and can automatically be derived, as will be seen in Chapter 8. If agents are built by different vendors, then it is possible that $P_k(x|\pi_k(x))$ and $P_m(x|\pi_m(x))$ are inconsistent for a pair of agents A_k and A_m. This may occur when $\pi_k(x) = \pi_m(x)$. For instance, in Figure 6.23(c), $P_1(g|h)$ by A_1 may be inconsistent with $P_2(g|h)$ by A_2. Inconsistence may also occur when $\pi_k(x) \neq \pi_m(x)$. For instance, in Figure 6.23(c), A_0 may have $P_0(j = j_0|i = i_1, p = p_0) = 0.7$, whereas A_2 may specify $P_2(j = j_0|i = i_1) = 0$. We make the following basic assumption to integrate independently built agents into a coherent system.

Basic Assumption 6.5 *Within each agent's subdomain, a JPD is consistent with the agent's belief. For shared variables, a JPD supplements an agent's knowledge with others'.*

The key issue is to combine agents' beliefs on a shared variable x to arrive at a common belief. A simple but less desirable method is to adopt the belief of the agent that contains $\pi(x)$. A more reasonable method (Poole et al. [55]) interprets the distribution from each agent as obtained from equivalent sample data. For example, consider a distribution $P(v|y)$, where $v, y \in \{0, 1\}$:

$$P(v = 0|y = 0) = 0.4 \quad \text{and} \quad P(v = 0|y = 1) = 0.2.$$

It can be interpreted as

$$P(v = 0|y = 0) = 48/120 \quad \text{and} \quad P(v = 0|y = 1) = 24/120.$$

That is, $v = 0$ in 48 out of 120 cases, where $y = 0$. The equivalent sample size 120 encodes how confident the vendor is about the distribution. The distribution $P(x|\pi(x))$ can then be obtained from the combined data sample. The exact procedure to obtain the combined $P(x|\pi(x))$ is not as important to the current discussion. The essence is that it is always possible (at least the simple method mentioned above can be used) for agents to combine their beliefs for each shared x to arrive at $P(x|\pi(x))$. With $P(x|\pi(x))$ thus defined for each shared x, and defined independently by the corresponding vendor for each internal x, the Basic Assumption 6.5 is fulfilled.

We now introduce MSBNs and show that the representation is implied by the basic assumptions.

Definition 6.15 *An MSBN M is a triplet (V, G, P): $V = \cup_i V_i$ is the **total universe** where each V_i is a set of variables called a **subdomain**. $G = \sqcup_i G_i$ (a hypertree MSDAG) is the **structure** where nodes of each subgraph G_i are labeled by elements of V_i. Let x be a variable and $\pi(x)$ be all parents of x in G. For each x, exactly one of its occurrences (in a G_i containing $\{x\} \cup \pi(x)$) is assigned $P(x|\pi(x))$, and each occurrence in other subgraphs is assigned a uniform potential. $P = \Pi_i P_i$ is the **JPD**, where each P_i is the product of the potentials associated with nodes in G_i. Each triplet $S_i = (V_i, G_i, P_i)$ is called a **subnet** of M. Two subnets S_i and S_j are said to be **adjacent** if G_i and G_j are adjacent in the hypertree.*

As an example, Figure 6.26 shows an MSBN based on the hypertree MSDAG in Figure 6.23. Each $C()$ denotes a uniform potential. For simplicity, the subscript in each $P_i()$ and $C_j()$ has been omitted. For subnet $S_0 = (V_0, G_0, P_0)$, we have $V_0 = \{f, i, j, o, p\}$ and $P_0(V_0) = P(j|i, p)P(p|f)C(i|f)P(o|f)C(f)$. Note that $P(p|f)$ and $P(o|f)$ are specified by the vendor of A_0, whereas $P(j|i, p)$ may be the result of combining $P_0(j|i, p)$ from A_0, $P_2(j|i)$ from A_2, and $P_3(j)$ from A_3.

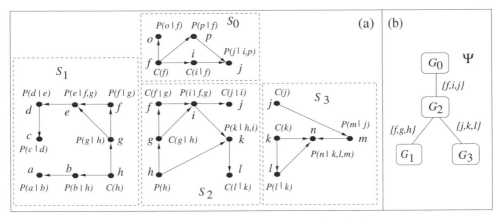

Figure 6.26: A simple MSBN. (a) Subnets. (b) The hypertree.

The JPD of the MSBN is

$$P(V) = P_0(V_0)P_1(V_1)P_2(V_2)P_3(V_3).$$

Given that for each variable $x \in V$, exactly one occurrence of x is assigned $P(x|\pi(x))$ and the uniform potential $C()$ assigned to other occurrences of x has no effect on the final product, we have

$$P_0(V_0)P_1(V_1)P_2(V_2)P_3(V_3) = \prod_{v \in V} P(v|\pi(v)),$$

which is equivalent to the JPD obtained in a single-agent Bayesian network (Theorem 2.3).

Note it is possible that, for some shared node x, there exist multiple local DAGs each of which contains $\pi(x)$. For example, in Figure 6.26, $\pi(f) = \{g\}$, $\pi(g) = \{h\}$, and $\pi(h) = \emptyset$. Both G_1 and G_2 contain each of these parent sets. In these cases, which occurrence of the shared node x is assigned $P(x|\pi(x))$ can be determined arbitrarily because the assignment makes no difference to the resultant JPD. In Figure 6.26, the occurrence of f in G_1 is assigned $P(f|g)$, the occurrence of g in G_1 is assigned $P(g|h)$, and the occurrence of h in G_2 is assigned $P(h)$.

Combining Proposition 6.14 and Basic Assumption 6.5, Theorem 6.16 establishes that, if one is committed to the basic assumptions, the representation of an MSBN must be adopted.

Theorem 6.16 *Let a multiagent system be one that observes Basic Assumptions 6.1 through 6.5. Then it is represented as an MSBN or equivalent.*

Proof: By Definition 6.15, an MSBN is defined by a hypertree MSDAG structure and a JPD. By Proposition 6.14, the hypertree MSDAG is implied by Basic

Assumptions 6.1 through 6.4. Hence, it suffices to show that Basic Assumptions 6.1 through 6.5 imply the JPD as defined in Definition 6.15.

Assume that for each variable x internal to a subgraph, $P(v|\pi(v))$ is specified by the corresponding agent and for each shared variable x, $P(v|\pi(v))$ is specified by combining beliefs from all agents involved. Then Basic Assumption 6.5 requires that the JPD of the universe $P'(V)$ satisfies

$$P'(V) = \prod_{v \in V} P(v|\pi(v)),$$

where $\pi(x)$ denotes the parents of x in G. According to Definition 6.15,

$$P(V) = \prod_i \prod_{v \in V_i} f(v|\pi_i(v)),$$

where $\pi_i(x)$ denotes the parents of x in G_i, $f(v|\pi_i(v)) = P(v|\pi_i(v))$ if $\pi_i(v) = \pi(v)$ and the occurrence of x in G_i is assigned $P(v|\pi(v))$, and $f(v|\pi_i(v))$, is a uniform potential otherwise. Because a one-to-one mapping exists between terms $P(v|\pi(v))$ in $P'(V)$ and nonuniform-potential terms in $P(V)$, and the uniform potentials do not affect the value of $P(V)$, we have $P'(V) = P(V)$.   □

As another example, consider the multiagent system for monitoring the digital system. Figures 6.27 through 6.31 show the subnet structure of each agent in the MSBN representation based on the virtual components U_0 through U_4 defined in Section 6.3. The figures were generated using the WebWeavr toolkit. Each variable in the subnet is labeled by its variable name followed by the variable's index (which readers may ignore) separated by a comma. The hypertree of the MSDAG is isomorphic to Figure 6.21(a).

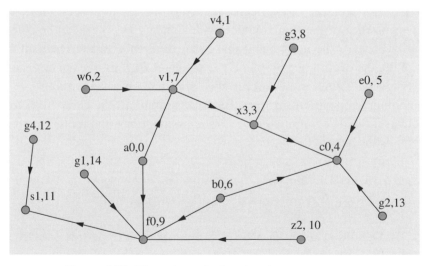

Figure 6.27: The subnet G_0 for virtual component U_0.

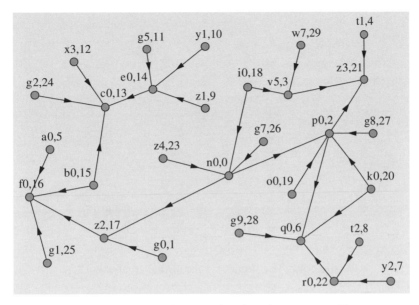

Figure 6.28: The subnet G_1 for virtual component U_1.

6.7 Bibliographical Notes

6.7.1 On Multiagent Systems

The field of multiagent systems, started about two decades ago under the banner of *distributed artificial intelligence* (DAI), has become very dynamic in both the

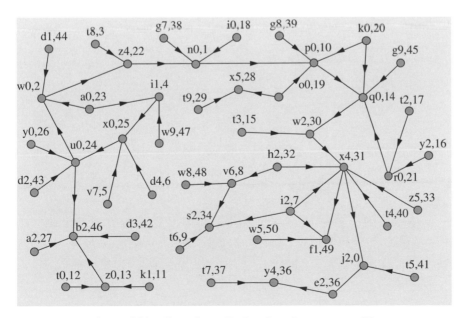

Figure 6.29: The subnet G_2 for virtual component U_2.

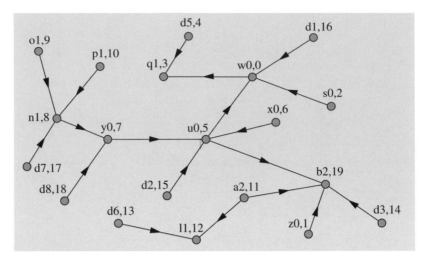

Figure 6.30: The subnet G_3 for virtual component U_3.

academic community and industry in recent years in response to the need to develop intelligent systems in large, complex, and open problem domains and the rapid development of the Internet and e-commerce. Numerous conferences are held each year such as the

- International Conference on Multiagent Systems (ICMAS),
- International Conference on Autonomous Agents,
- International Conference on Mobile Agents, and the
- International Conference on Cooperative Information Systems.

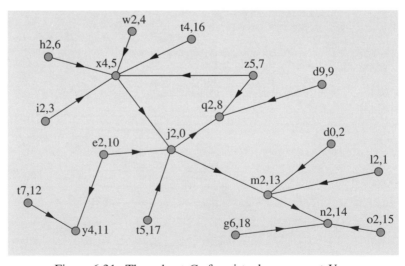

Figure 6.31: The subnet G_4 for virtual component U_4.

A classic collection of research articles in DAI is Bond and Gasser [5]. A more up-to-date introduction to the field of multiagent systems is provided by Weiss [77]. The first journal dedicated in the field is *Autonomous Agents and Multi-Agent Systems*, although several other journals publish research and advances in the field regularly such as the following:

- *Artificial Intelligence*,
- *IEEE Transactions on Systems, Man, and Cybernetics*, and
- *AI Magazine*.

6.7.2 On MSBNs

The research on MSBNs started in 1990 as the backbone of the PainULim project (Xiang et al. [94] and Xiang et al. [95]) conducted through collaboration between the University of British Columbia and Vancouver General Hospital. The practical goal of the project was to develop a prototype system for diagnosis of patients suffering from a painful or impaired upper limb due to diseases of the spinal cord, peripheral nervous system, or both. The scientific goal of the project was to study issues in developing Bayesian–network-based decision support systems in large and complex uncertain domains.

To allow a natural and modular representation of medical domain knowledge in neuromuscular diagnosis as well as focused and effective inference computation, the framework of MSBNs was proposed (Xiang [80]) initially under the single-agent paradigm. In this framework, subnets for natural subdomains can be constructed individually, and inference computation can be performed on one subnet at a time as the diagnosing neurologist shifts his or her attention from one subdomain to another.

At the 1994 UAI conference, Srinivas [69] independently presented his work on hierarchical model-based diagnosis with Bayesian networks. Careful examination revealed (Xiang [83]) that it was in fact an application of a special case of MSBNs to hierarchical model-based diagnosis. Later, Koller and Pfeffer [35] advanced the approach by Srinivas, proposed *object-oriented Bayesian networks* (OOBNs), and showed that MSBNs can be applied to inference in OOBNs.

At the same time, the modularity in MSBN representation and inference and the increasing awareness of the importance of multiagent systems inspired rethinking of MSBNs under the multiagent paradigm (Xiang [81] and Xiang [82]). The marginal probability distribution at each subnet and the JPD of the MSBN were given a multiagent interpretation. The sequential entering of observations and sequential local inference at centrally located subnets were extended to simultaneous local observations and asynchronous local inference by multiple distributed agents (Xiang [84]).

With the multiagent extension, MSBNs provide an effective framework for exact and distributed belief updating in cooperative multiagent systems. However, the representational restrictions of the framework were sometime questioned by researchers on multiagent systems. For instance, while I was presenting a demonstration on equipment monitoring with MSBNs at the First International Workshop on Multi-agent Systems at MIT's Endicott House in 1997, some participants questioned the necessity of the hypertree constraint for agent organization. In 1998, one of my colleagues at Aalborg University raised the issue of the necessity of the d-sepset constraint while I was visiting Denmark. The underlying fundamental question concerns the possibility of relaxing the representational constraints while preserving the desirable properties of MSBNs. The result of this inquiry first appeared in Xiang and Lesser [92] and formed the basis of this chapter.

6.8 Exercises

1. In Section 6.3 it was assumed that, for any two agents, natural or computational, to communicate directly, they must share some variables. Discuss the validity of this assumption.

2. Discuss the conditions under which a non-d-sepset agent interface is sufficient for e-message passing between the relevant agents.

3. The figure below shows the sectioning of a graph G (not shown) into a set of subgraphs. Determine if there exists a hypertree over G given the sectioning.

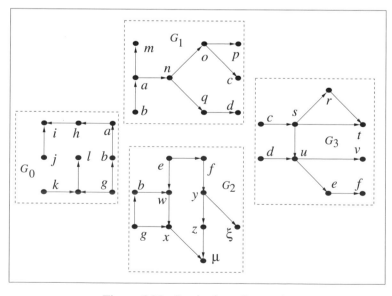

Figure 6.32: Sectioning of a graph.

4. The hypertree shown in Figure 6.24(a) and (c) does not satisfy the d-sepset condition. What is the minimum change to make in Figure 6.24(c) so that the resultant hypertree will satisfy the d-sepset condition?

5. Let $V = \cup_i V_i$ be a total universe in which each V_i is a subdomain and is associated with an agent A_i. Let $G = \sqcup_i G_i$ be a hypertree MSDAG in which nodes of each local DAG G_i are labeled by elements of V_i. For each node x, if $\pi(x) \subset V_i$, then A_i is responsible for acquiring $P(x|\pi(x))$ and ties are broken arbitrarily.

    Show that if G is an I-map of the total universe, then for each node x and each subdomain V_k, $P(x|\pi_k(x))$ is implied, where $\pi_k(x)$ denotes the parents of x in G_k.

6. Construct a digital system with five components using AND, OR, and NOT gates such that the components are interfaced as in Figure 6.5. Each component should have no less than five gates. Represent the system as an MSBN in which each subnet models a physical component.

7. Suppose five agents populate the digital system domain that you constructed in Exercise 6. Each agent needs to be informed about an additional gate in an interfacing component and its input and output. Define five virtual components accordingly and represent the system as an MSBN in which each subnet models a virtual component.

7

Linked Junction Forests

In Chapter 6, MSBNs were derived as the knowledge representation for multiagent uncertain reasoning under the five basic assumptions. As in the case of single-agent BNs, we want agents organized into an MSBN to perform exact inference effectively by concise message passing. Chapter 4 discussed converting or compiling a multiply connected BN into a junction tree model to perform belief updating by message passing. Because each subnet in an MSBN is multiply connected in general, a similar compilation is needed to perform belief updating in an MSBN by message passing. In this chapter, we present the issues and algorithms for the structural compilation of an MSBN. The outcome of the compilation is an alternative dependence structure called a *linked junction forest*. Most steps involved in compiling an MSBN are somewhat parallel to those used in compiling a BN such as moralization, triangulation, and junction tree construction, although additional issues must be dealt with.

The motivations for distributed compilation are discussed in Section 7.2. Section 7.3 presents algorithms for multiagent distributive compilation of the MSBN structure into its moral graph structure. Sections 7.4 and 7.5 introduce an alternative representation of the agent interface called a *linkage tree*, which is used to support concise interagent message passing. The need to construct linkage trees imposes additional constraints when the moral graph structure is triangulated into the chordal graph structure. Section 7.6 develops algorithms for multiagent distributive triangulation subject to these constraints. After triangulation, each agent has compiled its subnet into a local chordal graph structure. The compilation of the local chordal graph into a local junction tree plus several linkage trees is presented in Section 7.7. The resultant compiled structure for the multiagent system is a linked junction forest. Belief updating in the linked junction forest will be addressed in Chapter 8.

7.1 Guide to Chapter 7

Recall from Chapter 4 that, to perform belief updating by concise message passing in a nontree BN, the BN structure is converted into its moral graph, the moral graph is then triangulated, and finally the resultant chordal graph is organized into a JT. The moral graph, the chordal graph, and the JT are each produced in such a way as to preserve the dependence relations in the BN. That is, each of them is an I-map. A similar process is needed to compile the structure of an MSBN, its hypertree MSDAG, so that belief updating can be performed by multiple agents through concise message passing. The motivations for cooperative compilation by multiple agents, where each agent only has the access of its local subnet and no agent has the perspective of the entire MSBN, are addressed in Section 7.2.

Section 7.3 presents the first step of the compilation, moralization, which we illustrate in Figure 7.1. Three local DAGs, G_i ($i = 0, 1, 2$), of an (trivial) MSBN are shown in (a). The hypertree in this case is a hyperchain with G_2 at the center and G_0 and G_1 in terminal positions. Each local DAG is accessed by an agent A_i. Note that A_0 and A_2 share variables a, b, c, d and A_1 and A_2 share a, b, d.

The moralization can be initiated at any agent, say, A_2, which first performs a local moralization on G_2. The links $\langle a, c \rangle$ and $\langle c, d \rangle$ are added, and the directions of arcs are dropped. The resultant local moral graph is shown in (b). Agent A_2 then calls one of the adjacent agents, say, A_0.

In response, A_0 performs a local moralization on G_0. The links $\langle a, b \rangle$ and $\langle c, d \rangle$ are added, and the resultant local moral graph is shown in (c). After local moralization, A_0 sends to A_2 the added links among nodes shared with A_2, namely the links

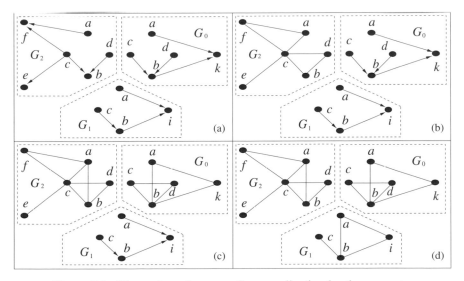

Figure 7.1: Illustration of cooperative moralization by three agents.

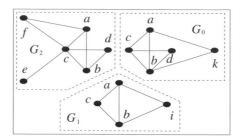

Figure 7.2: Result of cooperative moralization by three agents.

$\langle a, b \rangle$ and $\langle c, d \rangle$. Agent A_2 receives these links and adds them to its local graph if the links are not already included. In this case, the link $\langle a, b \rangle$ is added to G_2, as shown in (c). Afterwards, A_2 calls the other adjacent agent A_1.

In response, A_1 performs a local moralization on G_1. The link $\langle a, b \rangle$ is added, and the resultant local moral graph is shown in (d). Because $\langle a, b \rangle$ is between nodes shared with A_2, A_0 sends $\langle a, b \rangle$ to A_2. Inasmuch as the link is already in the local moral graph of A_2, no change will be made by A_2 when the link is received. In general, however, A_2 may need to add some links received from A_1.

Now each agent has performed the local moralization, and A_0 and A_1 have sent to A_2 added links among nodes shared with A_2. Next, it's A_2's turn to send added links among shared nodes. First, A_2 sends A_0 the added links among nodes shared with A_0, except those that A_0 sent to A_2 earlier. Hence $\langle a, c \rangle$ is the only link to send. When received, A_0 adds $\langle a, c \rangle$ to its local moral graph, and the result is shown in Figure 7.2. Similarly, A_2 sends $\langle a, c \rangle$ to A_1, which adds to its local moral graph. The end result of cooperative moralization for each agent is shown in Figure 7.2. The result is identical to a centralized moralization (Exercise 1). The algorithms for cooperative moralization in a general hypertree MSBN are presented in Section 7.3.

Once the moral graphs of local DAGs are obtained, they need to be triangulated so that each can be compiled into a JT representation. Section 7.4 describes how the resultant local JTs can support effective interagent communication through a derived data structure called a *linkage tree*. Figure 7.3 illustrates a linkage tree. The local moral graphs of agents A_0 and A_2 in the preceding example are duplicated in Figure 7.3(a). Each is a chordal graph, and the corresponding JT is shown in (b). The linkage tree L for communicating between the two agents is shown in (c). It is a cluster tree. Each cluster in L, called a *linkage*, has a corresponding cluster in each of the two local JTs, T_0 and T_2. The linkage provides a communications channel between them. Using the linkage tree, the belief of an agent over the agent interface $\{a, b, c, d\}$ can be decomposed into the belief over individual linkages. Suppose that each variable in the agent interface is ternary. An agent's belief over the interface then has a cardinality of $3^4 = 81$. Using the linkage tree, the belief

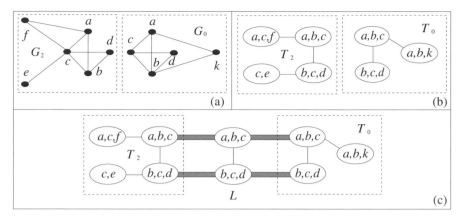

Figure 7.3: (a) Two local moral graphs. (b) Local JTs compiled from (b). (c) Linkage tree constructed from (b).

over $\{a, b, c\}$ has a cardinality of $3^3 = 27$ and so is the belief over $\{b, c, d\}$. Hence, the e-message encoded using the linkage tree has a cardinality of 54 and is more concise than 81. The general definition of the linkage tree is given in Section 7.4.

Section 7.5 reveals several important properties of linkage trees. We show that a linkage tree is also a junction tree and preserves the dependence relations among shared variables. These properties legitimate the decomposition of agent belief over the interface, as described in the preceding paragraph. Section 7.5 also shows that a linkage tree can be constructed if and only if the local moral graph is triangulated in a certain way. This imposes a requirement for cooperative triangulation to be presented in Section 7.6.

Section 7.6 develops the algorithms for cooperative triangulation of the local moral graphs. The technical requirements to be satisfied by the algorithms are enumerated in Section 7.6.1. Sections 7.6.2 through 7.6.5 develop the triangulation algorithms stepwise for the two-agent case, the hyperstar case (where a single agent is at the center and each other agent is adjacent to the center agent only), and the general hypertree case. The cooperative triangulation has a form similar to that of the cooperative moralization. It consists of local triangulations by individual agents and the exchange of fill-ins between adjacent agents. Section 7.6.6 proves that the algorithms satisfy all the technical requirements and are efficient. Some nontrivial examples are given to demonstrate the performance.

Section 7.7 discusses the construction of local JTs and linkage trees after cooperative triangulation. The construction of local JTs is similar to what is presented in Section 4.8. We describe the construction of the linkage tree between agents A_0 and A_2 for the preceding example (see Figure 7.3). Consider the local JT T_2 of agent A_2, which shares $\{a, b, c, d\}$ with A_0 and has "private" variables $\{e, f\}$. For

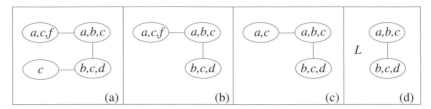

Figure 7.4: Illustration of linkage tree construction.

each private variable, if it is contained in a unique cluster, then A_2 removes it from the cluster. Figure 7.4(a) shows the resultant JT after e is removed from the cluster $\{c, e\}$. The new cluster $\{c\}$ is now a subset of the adjacent cluster $\{b, c, d\}$. It is then "merged" into $\{b, c, d\}$, as shown in (b). In (c), the other private variable f is removed because it is contained in a unique cluster. The new cluster $\{a, c\}$ is now a subset of the adjacent cluster and is merged. The resultant JT is shown in (d), which is the final linkage tree between A_2 and A_0.

7.2 Multiagent Distributed Compilation of MSBNs

An MSBN is a knowledge representation for distributed multiagent uncertain rea-soning. As in the single-agent paradigm, we want to perform exact inference ef-fectively by concise message passing. The message passing includes i-message passing within an agent for local inference as well as e-message passing between agents so that each can benefit from information maintained by others. We assume that local inference is a more frequent activity of an agent than communication with other agents. The analysis in Chapters 2 through 5 shows that, to make the local inference effective, each subnet should be compiled into a junction tree rep-resentation. Most steps involved in the MSBN compilation parallel those in the BN compilation. However, to perform communication effectively in an MSBN system, these steps are subject to further constraints and additional steps are needed, as will be seen in this chapter.

Should the compilation be performed centrally or distributively? We argue for the distributed compilation as follows: As elaborated in Basic Assumption 6.2, Propo-sition 6.3, and Definition 6.8, an agent can directly communicate only its belief on shared variables with an adjacent agent on the hypertree. Such a restriction is neces-sitated by *efficiency* of communication because it minimizes e-messages, as argued in Section 6.3. Additional advantages of the restriction can also be argued as follows:

1. The restriction protects the *privacy* of agents because they will not be required to reveal the unshared variables, the dependence among these variables, and the belief over these

variables. As an agent embeds the know-how of its vendor, and ultimately, the vendor's know-how is protected.

2. The restriction promotes *efficiency* in agent development. To build a member agent for a multiagent MSBN, only the d-sepsets with adjacent agents on the hypertree need to be coordinated. The agent vendor has the freedom to determine the internal characteristics of the subnet. Less coordination means better efficiency in development.

3. The restriction enhances *autonomy* of agents. Because only the d-sepsets with adjacent agents on the hypertree need to be coordinated, less coordination means more autonomy of each agent.

To preserve agent privacy, clearly it is desirable not to require subnets to be centralized and then compiled. Hence, we will concentrate on compilation algorithms that reveal no internal details of a subnet beyond the d-sepset.

An additional benefit of distributed compilation is that it facilitates dynamic formation of a multiagent MSBN. With dynamic formation, agents may join or leave the MSBN as the system is functioning. Recompilation may be necessary when the system structure changes. Distributed compilation requires no collection and redistribution of a large chunk of information (from and to diverse locations) and will be more efficient to deploy than its centralized counterpart.

Throughout the chapter, we consider an MSBN whose hypertree is populated by a set of agents. Each agent is located at a hypernode and maintains the corresponding subnet and d-sepsets. The hypertree structure plays an important role in supporting the compilation algorithms and ensuring their correctness. It provides a tree organization for agent interaction, a hypertree (Definition 6.8) decomposition of the total universe, and a hypertree decomposition of the DAG dependence structure. As we deal with different aspects of the compilation, we will switch from time to time among different aspects of the hypertree. When it is clear from the context, we will simply refer to the hypertree structure as *hypertree* and leave its relevant aspect implicit. During different activities, agents exchange different types of messages over the hypertree. We will refer to any message passing directly related to belief updating as *communication* and refer to other message passing activities, such as those during structure compilation, as *interaction* among agents. Just as the agent communication over the hypertree may be *direct* or *indirect*, the agent interaction may also be direct or indirect.

7.3 Multiagent Moralization of MSDAG

As in the single-agent paradigm, the first step in compilation is to convert the structure of an MSBN, namely, the hypertree MSDAG, into its moral graph. Recall from Definition 4.7 that the moral graph of a DAG is obtained by connecting parents

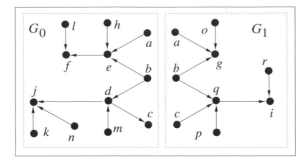

Figure 7.5: Moralization cannot be performed by local computation only.

pairwise for each child node and dropping the direction of each arc. Also recall from Theorem 4.8 that, given a minimal I-map in the form of a DAG, its moral graph is an undirected minimal I-map.

Formally, given a hypertree MSDAG $G = \sqcup_i G_i$, each subgraph G_i over V_i needs to be converted into a graph G'_i over V_i such that $\sqcup_i G'_i$ is the moral graph G' of G, and each pair of G'_i and G'_j is graph-consistent. The requirement of equality between $\sqcup_i G'_i$ and G' is one of maintaining minimal I-mapness. The requirement of graph-consistence demands consistence among agents with respect to their representation of dependence among shared variables. We refer to G'_i as the *local moral graph* and the collection of local moral graphs as the *moral graph* of the hypertree MSDAG. Due to the requirement of agent privacy as described in Section 7.2, the moralization computation should not disclose the internal details of each agent.

Local computation by individual agents without interaction does not ensure correct moralization of a hypertree MSDAG. Consider the two subnet structures in Figure 7.5. The moral graph of $G_0 \sqcup G_1$ contains a link $\langle b, c \rangle$ because b and c share the child q. However, if G_0 and G_1 are moralized independently by agents A_0 and A_1, A_0 would not include the link because b and c do not share any child in G_0.

We present recursive algorithms for each agent. The execution of each algorithm by an agent is activated by a call from an entity known as the *caller*. We denote the agent called to execute the algorithm by A_0. The caller is either an adjacent agent of A_0 in the hypertree or the system coordinator (a human or an agent). If the caller is an adjacent agent, we denote it by A_c. If A_0 has adjacent agents other than A_c, we denote them by $A_1, \ldots, A_n$. The subdomains of $A_c, A_0, \ldots, A_n$ are $V_c, V_0, \ldots, V_n$, and their subgraphs are $G_c, G_0, \ldots, G_n$, respectively. We denote $V_c \cap V_0$ by I_c and $V_0 \cap V_i$ by I_i ($i = 1, \ldots, n$).

We also present algorithms to be executed by the system coordinator. The system coordinator can be a human or a computational agent. In these algorithms, we use A_* to denote the agent selected by the system coordinator to perform certain operations.

In algorithm **CollectMlink**, an agent A_0 performs local moralization, updates its moral graph with links collected from adjacent agents, and then sends all relevant added moral links to the caller. Given a set F of links over a set X of nodes, we will call a subset $E \subseteq F$ a *restriction* of F to $S \subset X$ if

$$E = \{\langle x, y \rangle | x \in S, y \in S, \langle x, y \rangle \in F\}.$$

Algorithm 7.1 (CollectMlink) *Let A_0 be an agent with a local graph G_0 (initially a DAG). A **caller** is either an adjacent agent A_c or the system coordinator. Denote additional adjacent agents of A_0 by $A_1, \ldots, A_n$. When the caller calls on A_0, it does the following:*

set LINK $= \emptyset$;
moralize G_0 and denote added moral links by F_0;
add F_0 to LINK;
for each agent A_i $(i = 1, \ldots, n)$, do
    *call A_i to run **CollectMlink** and receive links F_i over I_i from A_i when finished;*
    add F_i to G_0 and LINK;
if caller is an agent A_c, send A_c the restriction of LINK to I_c;

After **CollectMlink** is finished in A_0, a set F_i of moral links received from each A_i is defined, and *LINK* contains links added from local moralization and all received links. Note that if a link is already in G_0 or *LINK*, then adding it has no effect.

In algorithm **DistributeMlink**, an agent A_0 receives moral links from the caller, updates its moral graph accordingly, and sends all relevant added moral links to each adjacent agent.

Algorithm 7.2 (DistributeMlink) *Let A_0 be an agent with a local undirected graph G_0 and a predefined set LINK of links. A **caller** is an adjacent agent A_c or the system coordinator. Denote additional adjacent agents of A_0 by $A_1, \ldots, A_n$. When the caller calls on A_0, it does the following:*

if caller is an agent A_c, do
    receive a set F_c of links over I_c from A_c;
    add F_c to G_0 and LINK;
for each agent A_i $(i = 1, \ldots, n)$, do
    send A_i the restriction of LINK to I_i with links in F_i removed;
    *call A_i to run **DistributeMlink**;*

Algorithm **CoMoralize** is executed by the system coordinator to activate cooperative distributive moralization.

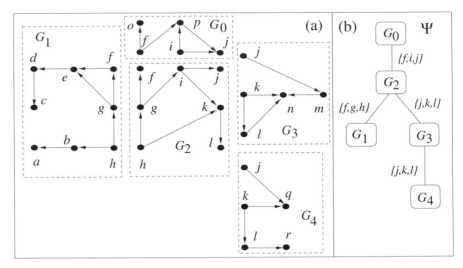

Figure 7.6: The subgraphs (a) and hypertree (b) of an MSBN.

Algorithm 7.3 (CoMoralize) *A hypertree MSDAG is populated by multiple agents with one at each hypernode. The system coordinator does the following:*

choose an agent A_* arbitrarily;
*call A_* to run **CollectMlink**;*
*after A_* has finished, call A_* to run **DistributeMlink**;*

As an example, consider the execution of **CoMoralize** in the MSBN shown in Figure 7.6. Suppose A_0 is selected as the A_*. In Figure 7.7(b), the black arrows

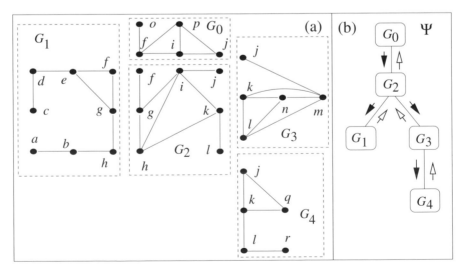

Figure 7.7: (a) Moral graphs after local moralization during **CollectMlink**. (b) During **CollectMlink**, activation follows black arrows and link passing follows white arrows.

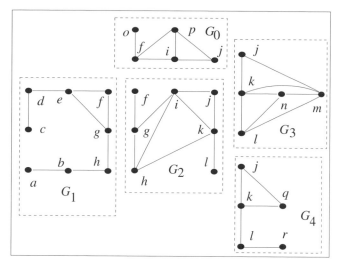

Figure 7.8: Moral graphs after **CollectMlink**.

show the direction of activation of **CollectMlink** through the hypertree. The moral graphs obtained after local moralization are shown in (a). Note that j and k are connected differently in G_2, G_3, and G_4 and f and i are connected differently in G_0 and G_2.

When A_4 finishes **CollectMlink**, it sends the link $\langle j, k \rangle$ to A_3. The white arrows in Figure 7.7(b) show the direction in which links will be sent (if any) between agents. When A_3 finishes, it in turn sends $\langle j, k \rangle$ to A_2. No link is sent from A_1 to A_2 or from A_2 to A_0. Hence, after **CollectMlink** is finished at A_0, the resultant local graphs are as shown in Figure 7.8.

Agent A_0 then executes **DistributeMlink**. It sends the link $\langle f, i \rangle$ to A_2 and call A_2 to run **DistributeMlink**. The black arrows in Figure 7.9(b) show the direction of activation of **DistributeMlink** through the hypertree. No more links are sent between other pairs of agents, and the final result is shown in Figure 7.9(a).

The following theorem shows that **CoMoralize** produces the same moral graph for the hypertree MSDAG as would be produced by a centralized moralization.

Theorem 7.1 *Let $G = \sqcup_i G_i$ be a hypertree MSDAG, G' be the moral graph of G, and G_i' be the local graph obtained by agent A_i after execution of **CoMoralize**. Then, each pair of G_i' and G_j' is graph-consistent and $G' = \sqcup_i G_i'$.*

Proof: Whatever links that an agent adds locally, it sends the relevant ones to each adjacent agents during **CollectMlink** and **DistributeMlink**. Hence, adjacent local graphs are graph-consistent. Because $G = \sqcup_i G_i$ is a hypertree MSDAG, any two local graphs are graph-consistent.

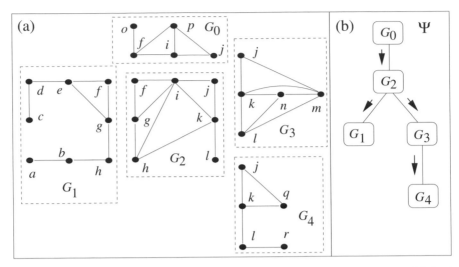

Figure 7.9: (a) Moral graphs after **DistributeMlink**. (b) During **DistributeMlink**, both activation and link passing follow black arrows.

Because G is a hypertree MSDAG, each non-d-sepnode has all its parents in its local subgraph, and each d-sepnode has all its parents in at least one local subgraph by Definition 6.11. Therefore, every node has all its parents in at least one local subgraph and the corresponding agent will add the necessary moral links among the parents as a centralized moralization would.   □

Another important feature of **CoMoralize** is that it does not reveal the internal details of each agent as required. The only information exchanged between adjacent agents are moral links that connect d-sepnodes. Hence, agent privacy is protected.

During **CoMoralize**, each of the n agents on the hypertree is called exactly twice. The local moralization dominates the computation at each agent. Each node in a subdomain is visited, and each pair of parents is processed. Let s denote the cardinality of the largest subdomain and m denote the cardinality of the largest parent set of individual variables. The complexity of local moralization is then $O(s\,m^2)$, and this is also the time complexity of **CoMoralize** for each agent. The overall complexity of **CoMoralize** for the entire multiagent system is then $O(n\,s\,m^2)$.

7.4 Effective Communication Using Linkage Trees

Each agent in an MSBN system needs to compile its subnet into a JT representation for effective local inference. After moralization, the next step is triangulation. Because the JT should also support communication, this need further constrains how the JT should be constructed. Since a given JT has a unique corresponding chordal graph, this need ultimately constrains how the moral graph should be triangulated. We consider such constraints with the following approach: Suppose that an agent has constructed a local JT from its subnet. We uncover desirable structural properties of this JT with respect to both local inference and communication. In view of the relation between a JT and its corresponding chordal graph, these properties become the constraints of triangulation.

To communicate between other agents, an agent needs to pass potentials (e-messages) over d-sepsets (Section 6.3). Given that the JT representation of a subnet maintains the current belief of the agent over its subdomain, how should the e-message over a d-sepset be obtained from the JT?

An obvious method is to construct the JT so that each d-sepset I with an adjacent agent is contained in a single cluster Q. The e-message will then be the potential $B(I)$ computed by marginalization of the cluster potential $B_Q(Q)$.

Consider two subnets whose structures are shown in Figure 7.10(a). After moralization, the local graphs are shown in (b), and they are chordal. From the cliques of the local graphs, the JTs in (c) can be constructed. However, no cluster in either JT contains the d-sepset $\{f, g, h\}$. This can be fixed by adding a link between f and h to each of the local graphs in (b). The resultant graphs are still chordal and convert to the JTs in (d). By performing local inference with these JTs, the potential over the d-sepset $\{f, g, h\}$ can be obtained directly from the corresponding clusters.

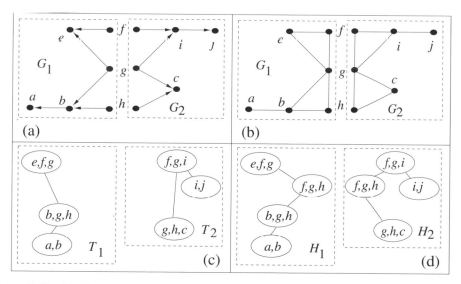

Figure 7.10: (a) The structures of two subnets. (b) Their local moral graphs. (c) JTs constructed from local moral graphs. (d) JTs constructed after adding link $\langle f, h \rangle$ to local moral graphs.

The price paid is the increased complexity in the JT representation. The complexity can be quantified in terms of the total number of potential values needed to specify the JT representation. We define the *raw-space complexity* of a JT representation to be

$$\sum_{Q} |B(Q)|,$$

where the summation is over all clusters of the JT. For example, if all variables are binary, the raw-space complexity of T_1 is 20 and so is that of T_2. Hence, the total raw-space complexity of T_1 and T_2 is 40, whereas that of H_1 and H_2 is 56.

In large problem domains, the d-sepsets will be much larger. The preceding method will increase raw-space complexity exponentially on the cardinality of d-sepsets. The consequences are increased space for storing the JTs and greater computational time for local inference at the same rate, not to mention the exponential complexity of e-message passing.

Is it possible to obtain the e-message from the less complex T_1 and T_2? The answer is yes. From the moral graphs in (b), $I(f, g, h)$ can be derived because they are I-maps. Therefore, an agent can compute the e-message

$$P(f, g, h) = P(f, g)P(g, h)/P(g),$$

where each factor is available from a cluster in T_1 and T_2. The graphical structure for computing the e-message is the cluster tree L shown in Figure 7.11. Each of $P(f, g)$ and $P(g, h)$ is a potential over a cluster of L, and $P(g)$ is the potential over the corresponding separator. Each cluster in L is connected to a cluster in T_1 and a cluster in T_2 by a shaded link. The link indicates from which cluster in the JT the corresponding cluster potential in L can be obtained. Next we present the idea above as a graphical structure called a *linkage tree* for effective computation of e-messages while keeping the raw-space complexity of the local JT low. First, we introduce a graphical operation called *merge* to be used in defining the linkage tree.

Definition 7.2 *A cluster C in a JT is **merged** into another cluster Q if Q is replaced by $Q' = Q \cup C$, C is removed from the JT (together with its separator with Q), and the other clusters originally adjacent to C are made adjacent to Q'.*

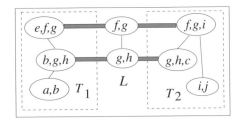

Figure 7.11: The cluster tree for computation of e-message.

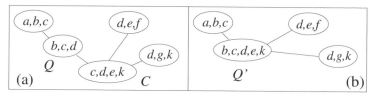

Figure 7.12: Cluster C in (a) is merged into Q, resulting in the cluster graph in (b).

An example is given in Figure 7.12. Using the merge operation, we define the linkage tree as follows:

Definition 7.3 *Let G be a subgraph in a hypertree MSDAG, I be the d-sepset between G and an adjacent subgraph, and T be a JT converted from G. Repeat the following procedure in T until no removal is possible:*

1. Remove $x \notin I$ if x is contained in a unique cluster C.
2. After removal, if C becomes a subset of an adjacent cluster D, merge C into D.

*Let L be the resultant cluster graph. Then L is a **linkage tree** of T with respect to I if*

$$\bigcup_{Q \text{ in } L} Q = I,$$

*where each cluster Q in L is called a **linkage**. A cluster in T that contains Q (breaking ties arbitrarily) is called the **linkage host** of Q.*

Consider T_1 in Figure 7.11, where $I = \{f, g, h\}$. Element $a \notin I$ is contained in a unique cluster $\{a, b\}$ and is removable. Afterwards, the cluster becomes a subset of the adjacent cluster $\{b, g, h\}$ and can be merged. The resultant cluster graph has two remaining clusters $\{b, g, h\}$ and $\{e, f, g\}$. Element b is removable and so is e. After removal of b and e, no more elements are removable. Because the generating set of the resultant cluster graph (shown as L) is I, L is a linkage tree of T_1. It has two linkages $\{g, h\}$ and $\{f, g\}$. The linkage host of $\{g, h\}$ is the cluster $\{b, g, h\}$ in T_1.

7.5 Linkage Trees as I-maps

We analyze the properties of linkage trees in relation to e-message passing and the constraints that they impose on triangulation. A linkage tree is constructed through the merge operation; hence, we analyze the property of this operation first. The following proposition shows that, after a cluster merge in a JT, the resultant cluster graph is still a JT.

Proposition 7.4 *Let C and Q be two adjacent clusters in a JT T. After C is merged to Q, the resultant cluster graph is still a JT.*

Proof: Denote the newly created cluster by Q' and the resultant cluster graph by T'. First, we show that T' is a tree. This is true because the merge operation is equivalent to deleting the link between C and Q and then joining the two resultant subtrees at the roots C and Q.

Next, we show that the intersection of each pair of clusters in T' is contained in every cluster on their path ρ. We consider the following possible cases:

1. ρ does not involve Q'.
2. ρ ends with Q'.
3. ρ has Q' in the middle.

For case 1, ρ is identical as in T and the statement is true on ρ.

For case 2, denote $\rho = \langle X, \dots, Y, Q' \rangle$. It suffices to show $X \cap Q' \subseteq Y$. Consider the path in T that starts from X, ends at C or Q, and contains both C and Q. Such a path is either $\langle X, \dots, Y, C, Q \rangle$ or $\langle X, \dots, Y, Q, C \rangle$, depending on whether X and Y are contained in the subtree rooted at C or at Q. For either case, because T is a JT, we have $X \cap Q \subseteq Y$ and $X \cap C \subseteq Y$. Hence, $X \cap Q' \subseteq Y$.

For case 3, denote $\rho = \langle X, \dots, Q', \dots, Y \rangle$. Consider the path $\rho' = \langle X, \dots, Y \rangle$ in T. The paths ρ and ρ' have the same clusters except C, and Q on ρ' is replaced by Q' on ρ. Because T is a JT, we have $X \cap Y \subseteq C$ and $X \cap Y \subseteq Q$. Hence, $X \cap Y \subseteq Q'$.   □

The next proposition says that a linkage tree is also a JT.

Proposition 7.5 *A linkage tree is a junction tree.*

Proof: After removal of an element contained in a unique cluster C of a JT, the junction tree property is unaffected. Hence, the resultant cluster graph is a JT. By Proposition 7.4, after merging a cluster into another, the resultant is still a JT.   □

Furthermore, the linkage tree preserves the I-mapness of the junction tree from which it was derived, as shown in Proposition 7.6.

Proposition 7.6 *Let L be a linkage tree of a JT T and I be the d-sepset. If T is an I-map over its generating set, then L is an I-map over I.*

Proof: We show that h-separation among elements in I portrayed by T is unchanged during the construction of L from T. That is, for any disjoint subsets X, Y, and Z of I, if $\langle X|Z|Y \rangle_T$, then $\langle X|Z|Y \rangle_L$.

Refer to Definition 7.3. Removal of x is irrelevant to h-separation among elements of I because x is not contained in any separator in T. Before C is merged into D, we have $C \subset D$. Hence, merging C into D does not alter h-separation along any path that includes C and D. □

Because a linkage tree L is a JT over the d-sepset I, the e-message can be obtained as

$$P(I) = \left[\prod_Q P(Q) \right] \Big/ \prod_S P(S),$$

where each Q is a linkage in L and each S is a separator in L. In the potential representation, this is equivalent to

$$B(I) = \left[\prod_Q B(Q) \right] \Big/ \prod_S B(S).$$

Clearly, as long as $|Q|$ is small, for each cluster the e-message can be efficiently encoded and passed even though $|I|$ is large.

So far, we have focused on the case in which

$$\bigcup_{Q \text{ in } L} Q = I$$

holds when the procedure in Definition 7.3 halts. Given a JT, it is possible that when the procedure halts we have

$$\bigcup_{Q \text{ in } L} Q \supset I,$$

in which case L is *not* a linkage tree. The following theorem identifies the condition under which the linkage tree exists. The condition says that the local chordal graph from which the JT is constructed must be triangulated in a certain way.

Before presenting the theorem, we revisit the notion of node elimination introduced in Section 4.6. There elimination was performed according to a total order. Here we extend the notion to partial orders. Let G be a graph over $N_0 \cup N_1$, where $N_0 \cap N_1 = \emptyset$. The graph G is eliminatable in the partial order (N_0, N_1) if it is possible to eliminate all nodes in N_0 one by one first and then eliminate all nodes in N_1 one by one without any fill-ins. Denote $j = |N_0|$ and $m = |N_1|$. A total elimination order $(v_0, v_1, \ldots, v_{j-1}, v_j, \ldots, v_{j+m-1})$ is consistent with a partial order (N_0, N_1) if $v_i \in N_0$ for all $i = 0, \ldots, j - 1$ and $v_i \in N_1$ for all $i = j, \ldots, j + m - 1$. That is, the first j elements in the total order are those in N_0 and the remaining are in N_1. The partial order could in general be $(N_0, N_1, \ldots, N_k)$. For simplicity, we write $(\{a\}, \{b\}, \{c, d\})$ as $(a, b, \{c, d\})$.

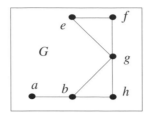

Figure 7.13: An eliminatable graph that is not eliminatable in the order ($\{a, g\}, \{b, e, f, h\}$).

Theorem 7.7 *Let G be a graph over V from which a JT T is constructed. Let I be a subset of nodes in G. Then a linkage tree of T exists with respect to I iff G is eliminatable in the order ($V \setminus I, I$).*

Proof: [Sufficiency] Suppose G is eliminatable in ($V \setminus I, I$). Consider a node $x \in V \setminus I$ that can be eliminated first without fill-ins. Then x must appear in a unique cluster C in T, for otherwise $adj(x)$ is incomplete. Hence, x can be removed from C. Repeating this argument for each node in $V \setminus I$, we will be able to remove all elements in $V \setminus I$ from T, and the resultant cluster graph corresponds to a linkage tree relative to I.

[Necessity] Suppose G is not eliminatable in ($V \setminus I, I$). That is, $V \setminus I$ cannot be eliminated without fill-ins in any total order that is consistent with ($V \setminus I, I$). This means that no matter what total order we use, there exists a nonempty subset of $V \setminus I$ (the subset may differ for different total orders) such that each node in the subset appears in at least two clusters of T. By Definition 7.3, elements in this subset will not be removed from T, and the generating set of the resultant cluster graph will be a proper superset of I. Hence, the linkage tree does not exist.   □

An eliminatable graph may not be eliminatable according to a given order. For example, the graph in Figure 7.13 can be eliminated in the order (a, b, e, f, g, h) but not in the partial order ($\{a, g\}, \{b, e, f, h\}$). To triangulate the graph using the order ($\{a, g\}, \{b, e, f, h\}$), a fill-in $\langle f, h \rangle$ must be added. Therefore, Theorem 7.7 shows that to allow an agent to communicate with an adjacent agent effectively using a linkage tree, the agent must triangulate its moral graph in the partial order ($V \setminus I, I$) before construction of its JT representation. Triangulation according to this constraint incurs possible overhead of additional fill-ins. In the next section, we consider the computation of multiagent triangulation.

7.6 Multiagent Triangulation

7.6.1 Problem Specification

After moralization, the next step is to triangulate the moral graph of the MSDAG into a chordal supergraph. This is necessary for the same reason that triangulation is used when compiling a BN: a junction tree of a graph exists if and only if the graph is chordal (Theorem 4.10). A moral graph is a minimal I-map (if the MSDAG is a minimal I-map). If it is not already chordal, to convert it into a chordal graph while maintaining the I-mapness, links must be added to obtain a chordal supergraph. We define the chordal graph from the MSDAG as the union of a set of local chordal graphs, each of which is a supergraph of a local moral graph. Once the local chordal graph is obtained, a JT I-map can be constructed from it for effective local inference.

Formally, given the moral graph $G = \sqcup_i G_i$ of a hypertree MSDAG, we need to convert each subgraph G_i over V_i into a chordal supergraph G'_i over V_i such that $\sqcup_i G'_i$ is a chordal supergraph G' of G (the chordality requirement) and each pair of G'_i and G'_j is graph-consistent (the *graph-consistence* requirement). As in multiagent moralization, the requirement of graph consistence is one of consistence in representing dependence relations among variables shared by multiple agents. By Theorem 7.7, for each G'_i and its d-sepset I_i relative to an adjacent subgraph on the hypertree, G'_i must be eliminatable in the order $(V_i \setminus I_i, I_i)$ (the *elimination order* requirement). Furthermore, for agent privacy, interaction between agents during triangulation should not reveal the internal dependency structure of each agent beyond the structure over the d-sepset (the *privacy* requirement).

Next we present algorithms for cooperative multiagent triangulation subject to the preceding four requirements. We proceed stepwise from the two-agent case to the hyperstar case and eventually to the general hypertree case.

7.6.2 Cooperative Triangulation by Two Agents

The simplest case of cooperative triangulation involves only two agents. Algorithm 7.4 performs the task.

Algorithm 7.4 *Let V_0 and V_1 be two subdomains that form a total universe and $I = V_0 \cap V_1 \neq \emptyset$. Let G_0 and G_1 be consistent graphs over V_0 and V_1 and be associated with agents A_0 and A_1, respectively.*

1. *A_0 eliminates nodes in G_0 in the order $(V_0 \setminus I, I)$; denote the fill-ins by F_0;*
2. *A_0 adds F_0 to G_0; denote the resultant graph by G'_0;*
3. *A_0 sends A_1 the restriction of F_0 to I;*
4. *A_1 adds to G_1 the restriction of F_0 to I received; denote the resultant graph by G'_1;*
5. *A_1 eliminates nodes in G'_1 in the order $(V_1 \setminus I, I)$; denote the fill-ins by F_1;*
6. *A_1 adds F_1 to G'_1; denote the resultant graph by G''_1;*
7. *A_1 sends A_0 the restriction of F_1 to I;*
8. *A_0 adds to G'_0 the restriction of F_1 to I received; denote the resultant graph by G''_0;*

Figure 7.14 illustrates the algorithm, and graphs G_0 and G_1 are shown in (a). The shared nodes are $I = \{a, b\}$. Elimination in G_0 is performed by agent A_0 in

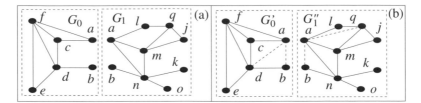

Figure 7.14: Illustration of Algorithm 7.4: (a) Before cooperative triangulation. (b) After cooperative triangulation.

the order $(\{e, f, c, d\}, \{a, b\})$. Node e can be eliminated without fill-in. No node in $\{f, c, d\}$ can then be eliminated without fill-ins. If f is eliminated next, a fill-in $\{a, d\}$ is required. Node c can then be eliminated without fill-in. To eliminate d next, another fill-in $\{a, b\}$ is needed. Hence, elimination in G_0 in the order (e, f, c, d, a, b) produces G_0' in (b), where the fill-ins are shown as dashed lines.

The fill-in $\{a, b\}$ is sent from A_0 to A_1 and added to G_1 to obtain G_1' (not shown). Elimination in G_1' by A_1 in the order $(k, o, j, l, q, m, n, a, b)$ produces fill-in $\{a, q\}$, and the resultant graph G_1'' is shown in Figure 7.14(b). For this example, A_1 has no fill-ins to send to A_0; hence, $G_0'' = G_0'$. However, this is not always the case.

The need for steps 7 and 8 in Algorithm 7.4 is illustrated in Figure 7.15. The graphs G_0 and G_1 are shown in (a), where $I = \{a, b, c, d\}$. Agent A_0 eliminates G_0 in the order (e, f, a, b, c, d), and obtains G', as shown in (b). After the restriction of fill-ins to I is sent to A_1, A_1 obtains Q' in (b). Agent A_1 starts elimination in Q' with k, which requires a fill-in $\{a, b\}$. This is the only fill-in added to obtain Q'' (not shown). Because $\{a, b\}$ is between nodes in I, A_1 sends it to A_0, which in turn adds it to G' to obtain G'' (not shown). Note that the last actions occur through the steps 7 and 8 of Algorithm 7.4.

The following proposition shows that when Algorithm 7.4 terminates, the local graphs are triangulated correctly with respect to all four requirements:

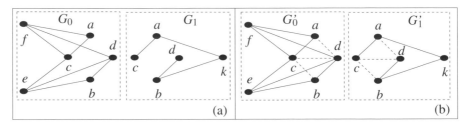

Figure 7.15: Illustration of the last two steps of Algorithm 7.4: (a) Before cooperative triangulation. (b) After cooperative triangulation.

Proposition 7.8 *When Algorithm 7.4 halts, the following hold:*

1. *Graphs G_0'' and G_1'' are graph-consistent.*
2. *Graph G_0'' is eliminatable in $(V_0 \backslash I, I)$ and G_1'' is eliminatable in $(V_1 \backslash I, I)$. Namely, each of G_0'' and G_1'' is chordal.*
3. *The graph union $G_0'' \sqcup G_1''$ is eliminatable in $((V_0 \cup V_1) \backslash I, I)$. Namely, $G_0'' \sqcup G_1''$ is chordal.*
4. *No internal characteristics of either local graph beyond links among I are revealed to the other agent.*

Proof:

(1) The statement is true because of steps 4 and 8.
(2) Graph G_1'' is eliminatable in the order $(V_1 \backslash I, I)$ from steps 5 and 6.
    Subset $V_0 \backslash I$ of nodes is eliminatable first from G_0'', because subgraphs of G_0'' and G_0' spanned by $V_0 \backslash I$ are identical by step 8, and G_0' is eliminatable in the order $(V_0 \backslash I, I)$ by steps 1 and 2. The remaining graph is spanned by I. It is eliminatable because G_1'' is eliminatable in the order $(V_1 \backslash I, I)$.
(3) It suffices to show that nodes in $G_0'' \sqcup G_1''$ are eliminatable in the order $(V_1 \backslash I, V_0 \backslash I, I)$. Subset $V_1 \backslash I$ is eliminatable, because the subgraph of $G_0'' \sqcup G_1''$ spanned by V_1 is G_1'' by steps 7 and 8, and G_1'' is eliminatable in the order $(V_1 \backslash I, I)$ from steps 5 and 6. The remaining graph is G_0'' by step 8 and is eliminatable in $(V_0 \backslash I, I)$, as shown in (2).
(4) Agent privacy is protected as the result of steps 3 and 7. □

Proposition 7.8 justifies the correctness of Algorithm 7.4. It is also needed for proving the more general case considered next.

7.6.3 Cooperative Triangulation in a Hyperstar

Next, we consider cooperative triangulation of the moral graph of an MSDAG when the hypertree is restricted to a hyperstar populated by $n > 2$ agents. In a hyperstar structure, exactly one agent is located at the center of the structure and each other agent is adjacent to the center agent only. We denote the agent at the center of the hyperstar by A_0 and the agent at each terminal by A_i $(i = 1, \ldots, n-1)$. We denote the local graph associated with agent A_i $(i = 0, \ldots, n-1)$ as G_i over V_i. Note that because the local graphs are derived from a hyperstar MSDAG, they satisfy the condition $V_i \cap V_j \subset V_0$ for every i and j $(i \neq j)$. We denote $V_0 \cap V_i$ by I_i.

Two algorithms are presented below. Algorithm 7.5 is executed by A_0 at the center, and Algorithm 7.6 is executed by each A_i $(i = 1, \ldots, n-1)$ at a terminal. Both algorithms are organized into two stages indicated by a blank line in between. It is assumed that if an existing link is added to a graph, there is no effect.

The triangulation starts by A_0. In each iteration of the first *for* loop (Algorithm 7.5), A_0 performs a local elimination relative to an I_i, adds fill-ins locally, and sends them to A_i. When A_i receives the fill-ins (Algorithm 7.6), it updates its local graph, performs a local elimination relative to I_i, adds fill-ins locally, and sends them to A_0 for inclusion. After this process is completed with respect to each A_i, A_0 finalizes G_0' and starts the second *for* loop (Algorithm 7.5). It sends all relevant fill-ins obtained in the first loop to each A_i. In response, each A_i receives the fill-ins and finalizes G_i' (Algorithm 7.6).

Algorithm 7.5 *Let A_0 be the agent at the center of the hyperstar. A_0 does the following:*

set LINK= ϕ;
for each terminal agent A_i, do
    eliminate V_0 in G_0 in the order $(V_0 \backslash I_i, I_i)$ and denote the resultant fill-ins by F;
    add F to G_0 and LINK;
    send A_i the restriction of LINK to I_i;
    receive a set F' of fill-ins over I_i from A_i;
    add F' to G_0 and LINK;
denote the resultant graph by G_0';

for each terminal agent A_i, do
    send A_i the restriction of LINK to I_i;

Algorithm 7.6 *Let A_i $(1 \le i \le n - 1)$ be an agent at a terminal of the hyperstar. A_i does the following:*

receive a set F of fill-ins over I_i from A_0;
add F to G_i;
eliminate V_i in G_i in the order $(V_i \backslash I_i, I_i)$ and denote the resultant fill-ins by F';
add F' to G_i;
send A_0 the restriction of F' to I_i;

receive a set LINK' of fill-ins over I_i from A_0;
add LINK' to G_i and denote the resultant graph by G_i';

Figure 7.16 illustrates Algorithms 7.5 and 7.6. The hyperstar with three agents is shown in (a), and the three local graphs are shown in (b). The d-sepsets are $\{a, b\}$ and $\{j, k\}$, respectively. After A_0 iterates once in the first *for* loop (relative to A_1) and A_1 completes its first stage, the resultant local graphs are those shown in (c). The corresponding elimination orders are $(o, k, j, l, q, m, n, a, b)$ and (e, f, c, d, a, b), respectively. After A_0 iterates the second time in the first *for* loop (relative to A_2) and A_2 completes its first stage, the resultant local graphs are those shown in (d). The

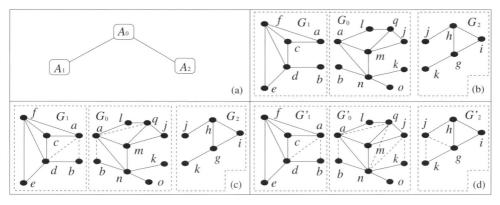

(a)

(b)

(c)

(d)

Figure 7.16: Illustration of Algorithms 7.5 and 7.6: (a) The hyperstar structure of a trivial MSBN. (b) The three local graphs. (c) The local graph after A_0 iterates once in the first *for* loop and A_1 completes its first stage. (d) The local graphs after A_0 iterates the second time in the first *for* loop and A_2 completes its first stage.

corresponding elimination orders are $(o, l, b, a, q, m, n, j, k)$ and (i, h, g, j, k), respectively. For this example, A_0 has no fill-ins to distribute in the second stage of Algorithm 7.5. This is not the case in the next example.

Consider the three local graphs in Figure 7.17(a) with the corresponding agents organized into the same hyperstar as in Figure 7.16(a). The d-sepsets are $\{a, b, d\}$

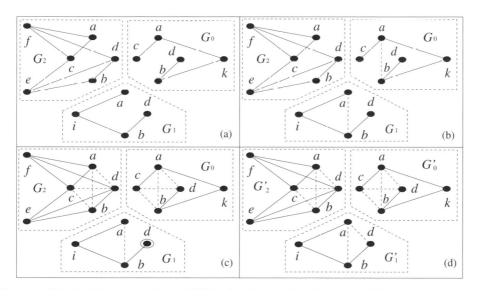

(a)

(b)

(c)

(d)

Figure 7.17: Need for second-round fill-in distribution in Algorithms 7.5 and 7.6. (a) Three local graphs of a trivial MSBN. (b) The local graphs after A_0 iterates once in the first *for* loop and A_1 finishes the first stage. (c) The local graphs after A_0 iterates the second time in the first *for* loop and A_2 finishes the first stage. (d) The local graphs after A_0 completes the second stage.

and $\{a, b, c, d\}$, respectively. After A_0 iterates once in the first *for* loop and A_1 finishes the first stage, the resultant graphs are those shown in (b). The elimination orders are (k, c, d, b, a) and (i, d, b, a), respectively. After A_0 iterates the second time in the first *for* loop and A_2 finishes the first stage, the resultant graphs are shown in (c). The elimination orders are (k, d, c, b, a) and (f, e, d, c, b, a), respectively. After A_0 completes the second stage, the local graphs in (d) are obtained. Note that without this stage, the fill-in $\{a, d\}$ cannot be added to G_1.

Also note that in the second stage, there is no effect when A_0 iterates through the terminal agent involved in the last iteration of the first *for* loop (i.e., A_2 in this example). This iteration is not excluded from Algorithm 7.5 to keep it simple.

Before trying to establish the correctness of Algorithms 7.5 and 7.6, we present a useful property of a chordal graph in Theorem 7.9, which says that the subgraph of a chordal graph is always chordal.

Theorem 7.9 *Let G be a chordal graph and G' be a graph obtained by deleting some nodes (and links incident to these nodes) from G. Then G' is chordal.*

Proof: It suffices to show that, after the deletion of a single node x, the remaining graph G' is chordal. We prove this by contradiction:

Suppose that G' is nonchordal. Then there must be a chordless cycle ρ of length greater than 3 in G'. Because G is chordal, a chord must have been deleted from ρ when x and links incident to it are deleted from G. That each link deleted has x as one of the end points implies that x is in ρ: a contradiction to the assumption that x is not in G'.   □

The following proposition shows that, when Algorithms 7.5 and 7.6 terminate, the local graphs are triangulated correctly with respect to the requirements on graph-consistence, chordality, and privacy. We will address the requirement on elimination order later.

Proposition 7.10 *Let Algorithms 7.5 and 7.6 be applied to the moral graph of a hyperstar MSDAG. When all agents halt, the following hold:*

1. *G'_0 and G'_i are graph-consistent for $i = 1, \ldots, n - 1$.*
2. *The graph $\sqcup_{i=0}^{n-1} G'_i$ is eliminatable in the order $(\cup_{i=1}^{n-1} (V_i \setminus I_i), V_0)$. Namely, $\sqcup_{i=0}^{n-1} G'_i$ is chordal.*
3. *G'_0 is chordal and G'_i is eliminatable in the order $(V_i \setminus I_i, I_i)$ for $i = 1, \ldots, n - 1$.*
4. *No internal details of any local graph beyond links among its d-sepsets are revealed to any other agent.*

Proof:

(1) For each i, G_0 and G_i are graph-consistent due to moralization. All fill-ins over I_i during the entire process are added to G_0', accumulated in *LINK*, and sent to A_i by A_0. They are added to G_i' by A_i. Hence, G_0' and G_i' are graph-consistent.

(2) Without losing generality, assume that the first *for* loop in Algorithm 7.5 proceeds in the order $i = 1, 2, \ldots, n - 1$. The eliminations then occurs in the following sequence:

$$A_0, A_1, A_0, A_2, \ldots, A_0, A_{n-1}.$$

Each A_i ($i = 1, \ldots, n - 1$) eliminates V_i in the order $(V_i \setminus I_i, I_i)$ (Algorithm 7.6). After the elimination, the subgraph spanned by $V_i \setminus I_i$ is never changed owing to hyperstar. Hence, $V_i \setminus I_i$ can be eliminated from G_i' without fill-ins.

The last two eliminations by A_0 and A_{n-1} are equivalent to Algorithm 7.4 due to hyperstar. By Proposition 7.8 (3), $G_0' \sqcup G_{n-1}'$ is eliminatable in the order

$$(V_0 \setminus I_{n-1} \cup V_{n-1} \setminus I_{n-1}, I_{n-1}),$$

which implies that G_0' is eliminatable in $(V_0 \setminus I_{n-1}, I_{n-1})$. Hence, all nodes in $\sqcup_{i=0}^{n-1} G_i'$ can be eliminated by first eliminating $V_i \setminus I_i$ for each $i = 1, \ldots, n - 1$ and then eliminating V_0.

(3) Consider the same elimination sequence $A_0, A_1, A_0, A_2, \ldots, A_0, A_{n-1}$ given in (2) above. For each G_i' ($i = 1, \ldots, n - 1$), $V_i \setminus I_i$ is eliminatable, as argued in (2) above; I_i is also eliminatable due to eliminability of G_0', the hyperstar, and Theorem 7.9.

(4) Agent privacy is protected because all fill-ins communicated are among the shared variables. □

Note that the statement "add F' to G_i" in Algorithm 7.6 is only included for clarity. It can be removed without affecting the end result.

7.6.4 On the Requirement of Elimination Order

The third statement in Proposition 7.10 is weaker than the requirement on elimination order, which would dictate that G_0' be eliminatable in the order $(V_0 \setminus I_i, I_i)$ for each $i = 1, \ldots, n - 1$. In fact, Algorithms 7.5 and 7.6 cannot guarantee this, as shown by the following example.

Consider the three local graphs in Figure 7.18(a), where G_0 is the center of the hyperstar. The two d-sepsets are $\{b, c, d, e\}$ and $\{a, d, e\}$, respectively. By Algorithms 7.5 and 7.6, A_0 first eliminates G_0 in the order (f, a, b, c, d, e) relative to A_1. The elimination produces no fill-ins. Then A_1 eliminates G_1 in the order (g, b, c, d, e) also without fill-ins.

Next, agent A_0 eliminates G_0 relative to A_2 in the order (f, b, c, a, d, e), which produces the fill-in $\{a, e\}$. The fill-in is added to G_0 and sent to A_2 (the dashed links

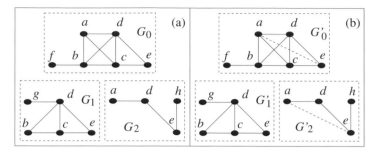

Figure 7.18: An example that does not satisfy the requirement on elimination order. (a) Before cooperative triangulation. (b) After algorithms and leaf are performed.

in (b)), which then eliminates G_2 in the order (h, a, d, e) relative to A_0 without additional fill-ins. All agents now halt, and each G_i' is shown in (b). However, if we try to eliminate the G_0' relative to A_1 in the order (f, a, b, c, d, e), another fill-in $\{b, e\}$ will be produced. Hence, G_0' is not eliminatable in the order $(V_0 \backslash I_1, I_1)$.

Such a situation does not seem to arise often. For instance, it does not occur in the examples in Figures 7.16 and 7.17. To guard against such situation, after Algorithms 7.5 has been terminated, A_0 may perform another round of elimination in the order $(V_0 \backslash I_i, I_i)$ for each $i = 1, \ldots, n - 2$. If any fill-in is added, the algorithm should be repeated. In Section 7.6.6, we revisit this issue in the more general case of cooperative triangulation in a hypertree.

7.6.5 Cooperative Triangulation in a Hypertree

We now consider the most general case: to triangulate the moral graph of a general hypertree MSDAG when the hypertree is populated by $n > 3$ agents. As in multi-agent moralization, we present recursive algorithms for each agent. The execution of each algorithm by an agent (denoted by A_0) is activated by a caller, which is either an adjacent agent (denoted by A_c) of A_0 or the system coordinator. If A_0 has adjacent agents other than A_c, denote them by $A_1, \ldots, A_m$. We denote $V_c \cap V_0$ by I_c and $V_0 \cap V_i$ by I_i ($i = 1, \ldots, m$). We also present algorithms to be executed by the system coordinator. The system coordinator can be a human or an agent. In these algorithms, we use A_* to denote the agent selected by the system coordinator to perform certain operations.

In algorithm **DepthFirstEliminate**, A_0 performs elimination and updating with respect to all adjacent agents. In algorithm **DistributeDlink**, A_0 receives fill-ins from the caller and then distributes all fill-ins among d-sepnodes added to G_0 since the start of the cooperative triangulation to each other adjacent agent. Algorithm **CoTriangulate** is executed by the system coordinator to activate the cooperative triangulation by multiple agents.

Algorithm 7.7 (DepthFirstEliminate) *Let A_0 be an agent with a local moral graph G_0. A **caller** is either an adjacent agent A_c or the system coordinator. Denote additional adjacent agents of A_0 by $A_1, \dots, A_m$. When the caller calls on A_0, it does the following:*

if caller is an agent A_c, do
  receive a set F_c of fill-ins over I_c from A_c;
  add F_c to G_0;

set LINK $= \phi$;
for each agent A_i ($i = 1, \dots, m$), do
  eliminate V_0 in the order $(V_0 \backslash I_i, I_i)$ and denote the resultant fill-ins by F;
  add F to G_0 and LINK;
  send A_i the restriction of LINK to I_i;
  *call A_i to run **DepthFirstEliminate** and receive fill-ins F' over I_i from A_i*
    when finished;
  add F' to G_0 and LINK;

if caller is an agent A_c, do
  eliminate V_0 in the order $(V_0 \backslash I_c, I_c)$ and denote the resultant fill-ins by F'_c;
  add F'_c to G_0 and LINK;
  send A_c the restriction of LINK to I_c;

Algorithm 7.8 (DistributeDlink) *Let A_0 be an agent with a local graph G_0. A **caller** is either an adjacent agent Λ_c or the system coordinator. Denote additional adjacent agents of A_0 by $A_1, \dots, A_m$. When the caller calls on A_0, it does the following:*

if caller is an agent A_c, do
  receive a set F_c of fill-ins over I_c from A_c;
  add F_c to G_0;

set LINK to the set of all fill-ins added to G_0 so far;
for each agent A_i ($i = 1, \dots, m$), do
  send A_i the restriction of LINK to I_i;
  *call A_i to run **DistributeDlink**;*

Algorithm 7.9 (CoTriangulate) *The moral graph of a hypertree MSDAG is populated by multiple agents with one at each hypernode. The system coordinator does the following:*

choose an agent A_* arbitrarily;
*call A_* to run **DepthFirstEliminate**;*
*after A_* has finished, call A_* to run **DistributeDlink**;*

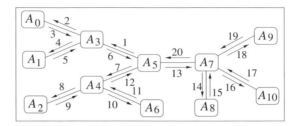

Figure 7.19: Illustration of **CoTriangulate**.

Figure 7.19 illustrates **CoTriangulate** with a system of 11 agents. The hypertree is depicted in the figure with each node labeled by an agent. Suppose the agent A_* chosen in the algorithm is A_5. For each arrow, an elimination (see **DepthFirstEliminate**) by the agent at the tail on its local graph is performed, and the relevant fill-ins generated are then sent to the agent at the head. For example, the arrow from A_5 to A_3 represents the elimination of A_5 on G_5 in the order $(V_5 \setminus V_3, V_5 \cap V_3)$ followed by sending A_3 the restriction of fill-ins to $V_5 \cap V_3$. The label of each arrow shows the sequence of the operation. It's easy to see that the sequence is similar to a depth-first traversal, which accounts for the name of the algorithm.

After A_5 has finished **DepthFirstEliminate**, the flow of fill-ins during execution of **DistributeDlink** is shown by *only* those arrows pointing away from A_5.

7.6.6 Correctness and Complexity of Cooperative Triangulation

In Theorem 7.12, we show the properties of **CoTriangulate**. We define the *altitude* of a node in a tree to be used in the proof. Consider a tree rooted at a node r. Given a node x on the tree, find the longest path from r through x to a terminal and denote the terminal by y. The altitude of x is then the length of the path between x and y. In Figure 7.19, for example, if the root is A_5, then the altitudes of A_2, A_4, and A_5 are 0, 1, and 2, respectively.

Lemma 7.11 establishes the depth-first property of **DepthFirstEliminate** to be used in proving Theorem 7.12. We will refer to A_* as the *root* of the hypertree.

Lemma 7.11 *All eliminations on local graphs located at the (hyper)subtree rooted at A_0 are performed after A_0 is called to run **DepthFirstEliminate** and before A_0 returns from the call.*

Proof: We prove by induction on the altitude k of A_0.

When $k = 0$, A_0 is a terminal, and exactly one elimination is performed on G_0 (see Figure 7.19, where the total number of eliminations on a local graph located at a hypernode is shown by the number of outgoing arrows of the node). Only the two *if* statements in **DepthFirstEliminate** are executed in this case, where the second one contains the elimination. Hence, the lemma is true.

Assume that the lemma is true when $k = q \geq 0$. Now consider the case $k = q + 1$. The eliminations performed on local graphs located at the subtree rooted at A_0 are those performed on G_0 and those performed on local graphs located at the subtree rooted at each A_i $(i = 1, \ldots, m)$. Exactly $m + 1$ eliminations are to be performed on G_0 (one relative to each A_i $(i = 1, \ldots, m)$ and one relative to A_c). The first m eliminations relative to A_i $(i = 1, \ldots, m)$ are contained in the *for* loop. All eliminations on local graphs located at the subtree rooted at each A_i $(i = 1, \ldots, m)$ are also performed in the *for* loop during the call to A_i by the inductive assumption. The last elimination on G_0 is performed in the *if* statement following the *for* loop. $\qquad\square$

We now prove the properties of **CoTriangulate**.

Theorem 7.12 *Let* **CoTriangulate** *be applied to the moral graph of a hypertree MSDAG. When all agents halt, the following hold:*

1. *Each pair of adjacent local graphs on the hypertree is graph-consistent.*
2. *The graph union of all local graphs on the hypertree is chordal.*
3. *Each local graph on the hypertree is chordal.*
4. *No internal details of any local graph beyond links among its d-sepsets are revealed to any other agent.*

Proof:

(1) Initially the moral graph is graph-consistent. We show that any fill-in among a pair of d-sepnodes produced by any agent will be communicated to any other agent that shares the pair. Let x and y be two disconnected d-sepnodes in G_0 associated with an agent A_0. Suppose the fill-in $\{x, y\}$ is produced by another agent A' during **Co-Triangulate**.

Because the MSDAG is a hypertree, nodes x and y are contained in every local graph on the path between A' and A_0. Suppose the path does not contain the root A_*. If A' has a lower altitude than A_0, then $\{x, y\}$ would be communicated to A_0 during **DepthFirstEliminate**. If A' has a higher altitude than A_0, $\{x, y\}$ would be communicated to A_0 during **DistributeDlink**.

Next suppose the path does contain the root A_*. Then $\{x, y\}$ would be communicated from A' to A_* during **DepthFirstEliminate** and from A_* to A_0 during **DistributeDlink**.

(2) We prove the statement by induction on the altitude k of agent A_0. When $k = 1$, A_1 through A_m are terminals of the hypertree. By Lemma 7.11, all eliminations on graphs located on the subtree rooted at A_0 are performed when A_0 is called to run **DepthFirstEliminate**. Afterwards, no change is made on each local subgraph spanned by $V_i \setminus I_i$. The elimination process is identical to that of Algorithms 7.5 and 7.6 except for the additional elimination of G_0 relative to A_c. Because the additional elimination is performed last and on V_0 only, by Proposition 7.10 (2), the graph union Q of G_0 through G_m is eliminatable in the order $(\cup_{i=1}^{m}(V_i \setminus I_i), V_0 \setminus I_c, I_c)$.

Denote the union of all local graphs on the subtree rooted at A_i as Q_i over V_i' ($i = 1, \ldots, m$). When the altitude of A_0 is $k = q \geq 1$, assume that Q_i is eliminatable in the order $(\cup_{i=1}^m (V_i' \backslash I_i), V_0 \backslash I_c, I_c)$.

Consider the case $k = q + 1$. By the inductive assumption, each Q_i is eliminatable in $(V_i' \backslash I_i, I_i)$. Because the last elimination on the subtree rooted at A_0 eliminates V_0 in $(V_0 \backslash I_c, I_c)$, the graph union of all local graphs on the subtree is eliminatable in the order $(\cup_{i=1}^m (V_i' \backslash I_i), V_0 \backslash I_c, I_c)$. Finally, when $A_0 = A_*$, the graph union of all local graphs on the subtree is eliminatable in the order $(\cup_{i=1}^m (V_i' \backslash I_i), V_0)$ by Proposition 7.10 (2).

(3) Each local graph G_0 other than the one associated with A_* is eliminated the last time in the order $(V_0 \backslash I_c, I_c)$. Hence, $V_0 \backslash I_c$ is eliminatable. By (2) and Theorem 7.9, I_c is eliminatable.

For the root A_*, the last elimination on G_0 is performed relative to an A_i in the order $(V_0 \backslash I_i, I_i)$. The processing is equivalent to Algorithm 7.4. By the hypertree and Proposition 7.8 (2), G_0 is eliminatable in $(V_0 \backslash I_i, I_i)$.

(4) All fill-ins communicated are among the shared variables. $\square$

Next, we consider the time complexity of **CoTriangulate**. We concentrate on **DepthFirstEliminate** and focus on its elimination processing only. This is because the amount of computation for transmission of fill-ins to adjacent agents during **DepthFirstEliminate** and **DistributeDlink** is dominated by the former.

Given the moral graph of a hypertree MSDAG with n hypernodes, it is easy to see from Figure 7.19 that $2(n - 1)$ eliminations are performed during **DepthFirstEliminate**. Let k be the maximum number of nodes in a local graph and d be the maximum degree of a node. To eliminate a node, the completeness of its adjacency is checked. The complexity of the checking is $O(d^2)$. Using the heuristics in Section 4.6, $O(k)$ nodes are checked before one is eliminated. Hence, the time complexity of eliminating all nodes in a local graph is $O(k^2 d^2)$. The complexity of **CoTriangulate** is then $O(n\, k^2\, d^2)$.

As shown in Section 7.6.4, **CoTriangulate** does not guarantee that each G_0 is eliminatable in $(V_0 \backslash I_i, I_i)$, although negative cases do not occur often. The following algorithm extends **CoTriangulate** to ensure the requirement on eliminatable order is fully satisfied.

Algorithm 7.10 (SafeCoTriangulate)

perform CoTriangulate;
each agent performs an elimination relative to the d-sepset with each adjacent
*   agent;*
if no agent added any fill-ins, halt;
else restart this algorithm;

The following theorem establishes the property of **SafeCoTriangulate**:

Table 7.1: *Agent communication interfaces*

Agents	Interface
A_0 vs A_1	$I_{0,1} = \{a, b, c, d, e, f, g\}$
A_1 vs A_2	$I_{1,2} = \{h, i, j, k, l, m, n, o, p, q, r\}$
A_2 vs A_3	$I_{2,3} = \{s, t, u, v, w, x, y, z, a', b'\}$
A_2 vs A_4	$I_{2,4} = \{c', d', e', f', g', h', i', j'\}$

Theorem 7.13 *Let **SafeCoTriangulate** be applied to the moral graph of a hypertree MSDAG. When all agents halt, each local graph G_0 is eliminatable with respect to each adjacent local graph G_i in the order $(V_0 \backslash I_i, I_i)$.*

Proof: Each round of **SafeCoTriangulate** will add some fill-ins to some local graphs, for otherwise it will halt. Because only a finite number of fill-ins can be added to each local graph before it becomes complete, and a complete graph is eliminatable in any order, **SafeCoTriangulate** will halt. □

By Theorems 7.12 and 7.13, we conclude that the problem of cooperative triangulation with respect to all requirements is solved by **SafeCoTriangulate**. Figure 7.20(a) shows the moral graph of a hypertree MSDAG whose total universe contains 80 variables. The hypertree is shown in Figure 7.21, and the agent interfaces are listed in Table 7.1.

The result of **SafeTriangulate** is shown in Figure 7.20(b) with fill-ins displayed as dashed lines. **SafeTriangulate** terminates with one execution of **CoTriangulate**. A total of eight fill-ins are added (identical fill-ins are counted only once). One might ask whether cooperative triangulation will result in too many fill-ins compared with a single-agent centralized triangulation. To answer this question empirically, an experiment has been performed in which **GetChordalGraph** in Section 4.6 was applied to the union of the local graphs in Figure 7.20(a). As a result, a total of six fill-ins were added. Hence, the cooperative triangulation results in only two extra fill-ins for this example. The experiment demonstrates that cooperative triangulation produces a reasonably sparse chordal graph.

In the following, we show the results of cooperative triangulation in the multi-agent system for monitoring the digital system presented in Chapter 6. Figure 7.22 shows the local chordal graph for agent A_0. Dark links are those in the original local DAG. Light links are added during moralization. The only fill-in is $\{a_0, x_3\}$ (the grey level of the link is intermediate).[1]

---

[1] The three types of links in black, light grey, and darker grey correspond to black, green, and red links on computer screen, respectively.

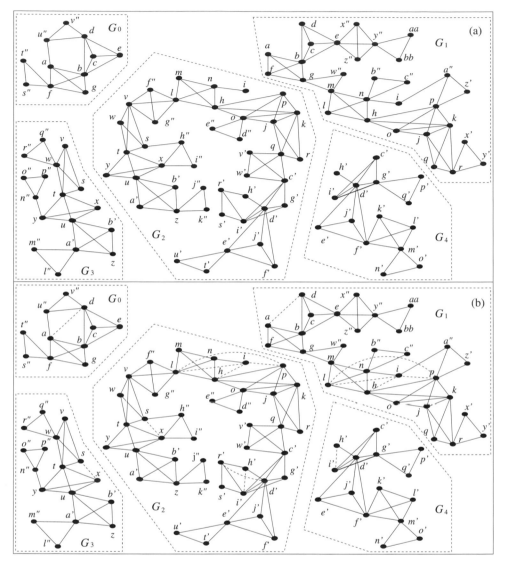

Figure 7.20: (a) The moral graph of a hypertree MSDAG. (b) The chordal graph obtained by **SafeTriangulate**.

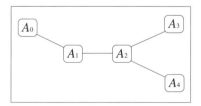

Figure 7.21: The hypertree of an MSDAG.

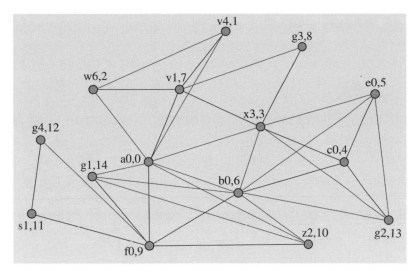

Figure 7.22: The local chordal graph for A_0.

Figure 7.23 shows the local chordal graph for A_1. The fill-in $\{a_0, x_3\}$ is shared with A_0. An additional fill-in $\{i_0, p_0\}$ is added. Figure 7.24 shows the local chordal graph for A_2. The fill-in $\{i_0, p_0\}$ is shared with A_1, and $\{s_0, x_0\}$ is added as an additional fill-in. Figures 7.25 and 7.26 show the local chordal graphs for A_3 and A_4. Agent A_3 shares $\{s_0, x_0\}$ with A_2, and A_4 has no fill-ins in its local graph.

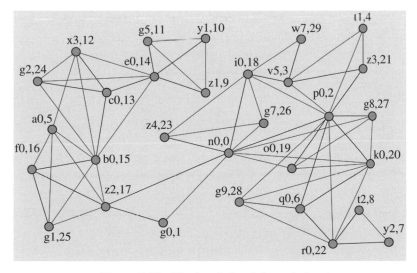

Figure 7.23: The local chordal graph for A_1.

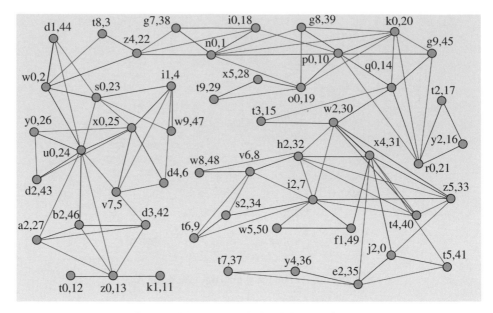

Figure 7.24: The local chordal graph for A_2.

7.7 Constructing Local Junction Trees and Linkage Trees

After each local dependence structure has been moralized and triangulated, it can be further organized into a local junction tree. This task can be performed by each individual agent using the method described in Section 4.8. After the local JT

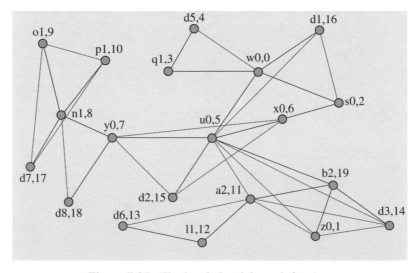

Figure 7.25: The local chordal graph for A_3.

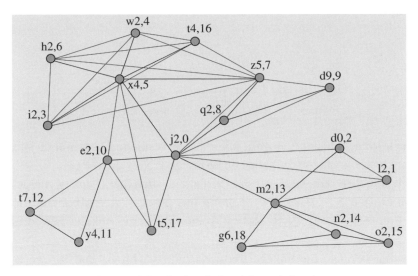

Figure 7.26: The local chordal graph for A_4.

is obtained, a linkage tree needs to be constructed between the agent and each adjacent agent. Because the local graph has been triangulated with respect to the requirement on elimination order, such a linkage tree does exist (Theorem 7.7) and can be constructed from the local JT according to Definition 7.3. We illustrate this process with an example.

Figure 7.27 shows the local JTs obtained by agents A_1 and A_2. Each can compute the local JT using its local chordal graph shown in Figure 7.20(b). No interaction is needed.

Figure 7.28 shows the linkage tree obtained by A_1 and A_2. Agent A_1 applies the procedure in Definition 7.3 to the local JT in Figure 7.27(a). It can remove aa and bb

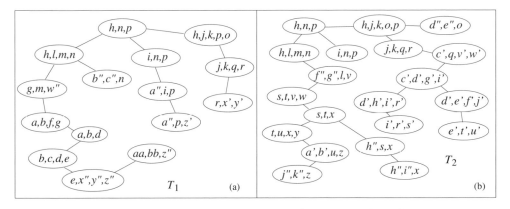

Figure 7.27: The local junction trees for A_1 (a) and A_2 (b).

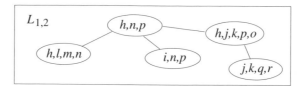

Figure 7.28: The linkage tree between A_1 and A_2.

from the terminal cluster $C = \{aa, bb, z''\}$. The resultant C becomes a subset to its adjacent cluster $\{e, x'', y'', z''\}$ and can be merged. Now denote $C = \{e, x'', y'', z''\}$. Agent A_1 can then remove x'', y'', and z'' from C such that C can be merged into its adjacent cluster $\{b, c, d, e\}$. By repeating this process, A_1 will obtain the linkage tree in Figure 7.28. Similarly, A_2 will obtain the same linkage tree locally from Figure 7.27(b).

Note that because the linkage trees associated with A_1 and A_2 are computed locally and independently; in general, the two linkage trees may be different. We will return to this issue in Section 8.3.

Before we continue, we show the local junction trees obtained by the multiagent system for monitoring the digital system in Figures 7.29 through 7.33.

From these local junction trees, the linkage trees between each pair of adjacent agents are computed, as shown in Figures 7.34 through 7.37. Note that although the d-sepset between a pair of agents has a cardinality between 9 and 13 (see Table 6.2), the largest linkage among all linkage trees has a cardinality of only 6.

With the local JT and linkage trees constructed by each agent, the hypertree MSDAG in the original MSBN has been converted into a different dependence structure, which is termed a *linked junction forest* (LJF). We formally define this structure as follows:

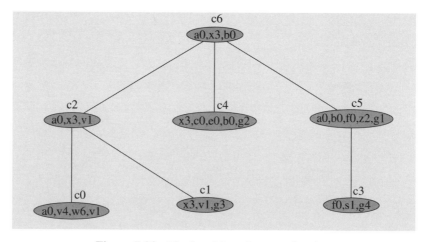

Figure 7.29: The local junction tree for A_0.

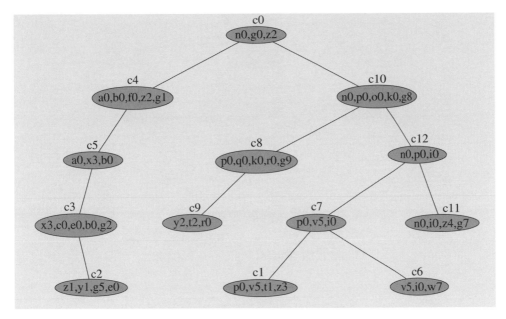

Figure 7.30: The local junction tree for A_1.

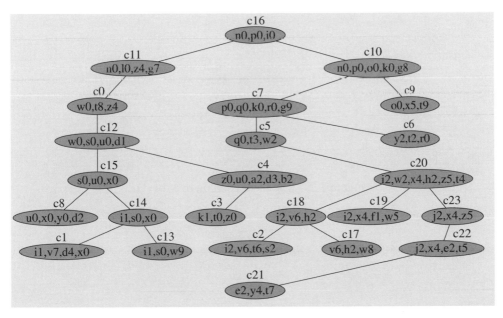

Figure 7.31: The local junction tree for A_2.

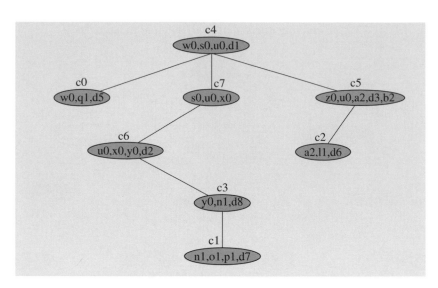

Figure 7.32: The local junction tree for A_3.

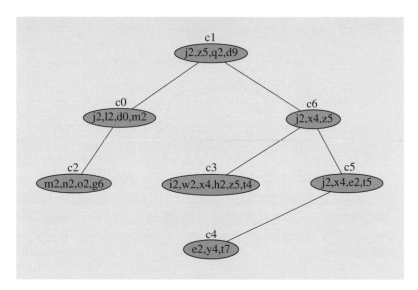

Figure 7.33: The local junction tree for A_4.

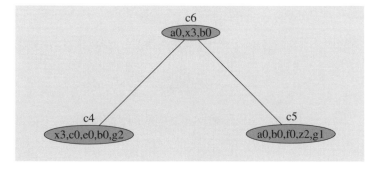

Figure 7.34: The linkage tree between A_0 and A_1.

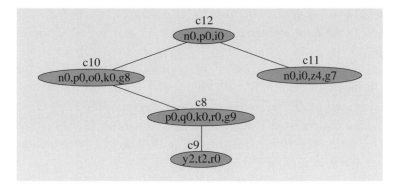

Figure 7.35: The linkage tree between A_1 and A_2.

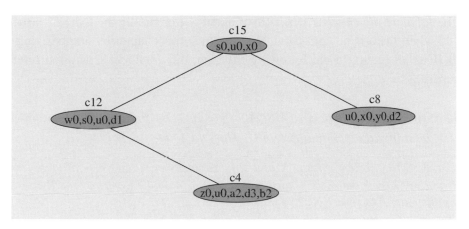

Figure 7.36: The linkage tree between A_2 and A_3.

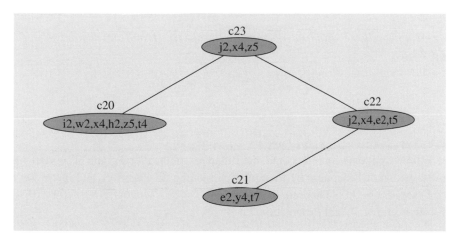

Figure 7.37: The linkage tree between A_2 and A_4.

Definition 7.14 *A linked junction forest F is a tuple (V, G, T, L).*

*$V = \cup_i V_i$ is the **total universe**, where each V_i is a set of variables, called a* **subdomain***.*

$G = \sqcup_i G_i$, where each $G_i = (V_i, E_i)$ is a chordal graph such that there exists a hypertree Ψ over G.

$T = \{T_i\}$ is a set of JTs, each of which is a corresponding JT of G_i.

$L = \{L_i\}$ is a collection of linkage tree sets. Each $L_i = \{L_{i,j}\}$ is a set of linkage trees, one for each hyperlink incident to G_i in Ψ. Each $L_{i,j}$ is a linkage tree of T_i with respect to a hyperlink $V_i \cap V_j$.

An LJF is an alternative dependence structure for a multiagent system, where each agent is associated with a (V_i, G_i, T_i, L_i). We refer to the tuple (V_i, G_i, T_i, L_i) as the *local JT dependence structure* of agent A_i, or simply the local JT structure of A_i when there is no confusion. We conclude this chapter by emphasizing that an LJF is an I-map. In the following theorem, the graphical separation refers to h-separation.

Theorem 7.15 *Let a hypertree MSDAG $G = \sqcup_i G_i$ be an I-map of a total universe and F be a linked junction forest of G. Then the following holds:*

1. *For each two disjoint subsets X and Y of variables in the universe, if there exists a hyperlink Z such that X is located on one side of Z on the hypertree and Y is located on the other side, then $I(X, Z, Y)$.*
2. *For each subdomain V_i with the corresponding JT T_i in F and any disjoint subsets X, Y, and Z of V_i,*

$$\langle X|Z|Y \rangle_{T_i} \Longrightarrow I(X, Z, Y).$$

3. *For each hyperlink I with a corresponding linkage tree L and any disjoint subsets X, Y, and Z of I,*

$$\langle X|Z|Y \rangle_L \Longrightarrow I(X, Z, Y).$$

Proof:
(1) It follows from Theorem 6.12 and the I-mapness of G.
(2) It follows from the I-mapness of G, the I-mapness of the moral graph (Theorem 4.8), the preservation of I-mapness by triangulation (adding fill-ins only), and the preservation of I-mapness by JTs (Theorem 4.18).
(3) It follows from (2) and Proposition 7.6.   □

Given an LJF $F = (V, G, T, L)$, if the three conditions in Theorem 7.15 hold, F is said to be an *I-map* over V.

Theorem 7.15 describes the three levels of conditional independence encoded in an LJF. The hypertree-level independence ensures effective and exact distributed inference, the JT-level independence ensures effective and exact local inference, and the linkage-tree level independence ensures effective and exact e-message passing. These will become even clearer as we present the MSBN inference methods in the next chapter.

7.8 Bibliographical Notes

The concept of linkage trees was initially presented in Xiang, Poole, and Beddoes [96] and was refined in Xiang [84, 89]. The analysis of the cluster merge operation (Proposition 7.4) is from Jensen [28]. The distributed triangulation of a hyperstar was first proposed in Xiang et al. [96]. The cooperative triangulation of a general hypertree was proposed by Xiang [89].

7.9 Exercises

1. Verify that the result of cooperative moralization in Figure 7.2 is identical to what would be obtained from a centralized moralization.
2. Follow Algorithm **CoMoralize** with the MSDAG in Figure 7.6 on the assumption that A_2 is selected as A_*. Compare the resultant moral graph of the MSDAG with that from Section 7.3.
3. Compute the linkage tree between agents A_2 and A_3 from the local junction tree of A_2 in Figure 7.31, where the local junction tree of A_3 is shown in Figure 7.32.
4. Apply two-agent triangulation (Algorithm 7.4) to the local moral graphs in Figure 7.10(b). Interpret the result.
5. Follow three-agent triangulation on the assumption that the local moral graphs are those in Figure 7.17(a). Let agent A_2 execute Algorithm 7.5 and A_0 and A_1 execute Algorithm 7.6.
6. Follow Algorithm **SafeCoTriangulate** on the assumption that the local moral graphs are those in Figure 7.18(a). Let A_0 be the initiating agent A_*.
7. Follow Algorithm **SafeCoTriangulate** on the assumption that the local moral graphs are those in Figure 7.20(a). Let A_1 be the initiating agent A_*.
8. Use the local chordal graphs in Figure 7.20(b) to perform the following:
   (a) Construct the local JTs T_0 and T_1 for agents A_0 and A_1.
   (b) Construct the linkage tree between A_0 and A_1 from T_0.
   (c) Construct the linkage tree between A_0 and A_1 from T_1.

8

Distributed Multiagent Inference

Chapter 7 has presented compilation of an MSBN into an LJF as an alternative dependence structure suitable for multiagent belief updating by concise message passing. Just as in the single-agent paradigm in which the conditional probability distributions of a BN are converted into potentials in a junction tree model, the conditional probability distributions in an MSBN need to be converted into potentials in the LJF before inference can take place. This chapter presents methods for performing such conversions and passing potentials as messages effectively among agents so that each agent can update belief correctly with respect to the observations made by all agents in the system.

Section 8.2 defines the potential associated with each component of an LJF and describes their initialization based on probability distributions in the original MSBN. Section 8.3 analyzes the topological structures of two linkage trees over an agent interface computed by two adjacent agents through distributed computation. This analysis demonstrates that, even though each linkage tree is created by one of the agents independently, the two linkage trees have equivalent topologies. This result ensures that the two agents will have the identical message structures when they communicate through the corresponding linkage trees. Sections 8.4 and 8.5 present direct interagent message passing between a pair of agents. The effects of such message passing are formally established. The algorithms for multiagent communication through intra- and interagent message passing are presented in Section 8.6. We also establish that exact posterior distributions can be obtained efficiently by local computation after such communications. Section 8.7 demonstrates how multiagent inference functions in trouble-shooting a digital system. The computational complexity of multiagent communication is analyzed in Section 8.8. The properties of, and motivations for regional agent communication are shown in Section 8.9. The possibility of extending the loop cutset conditioning and forward sampling (two alternative methods for belief updating in BNs) to inference in MSBNs is considered in Section 8.10.

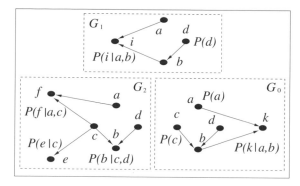

Figure 8.1: A trivial MSBN.

8.1 Guide to Chapter 8

Chapter 7 has described compiling an MSBN into an LJF structure. Before belief updating can be performed, the quantitative knowledge in the MSBN needs to be compiled as well. Section 8.2 presents the method for the compilation. As an example, the MSBN in Section 7.1 is duplicated in Figure 8.1 with the conditional probability distributions shown. The hypertree has the topology of $G_0 - G_2 - G_1$. Its compiled LJF representation is shown in Figure 8.2, where L_{02} is the linkage tree between T_0 and T_2 and L_{12} is the linkage tree between T_1 and T_2. The probability distribution of each node in each subset is assigned to a cluster in the corresponding local JT in the same way as in the single-agent paradigm. For example, $P(k|a, b)$ is associated with the node k in G_0. Because the cluster $\{a, b, k\}$ in the local *JT* T_0 contains the family of k, $P(k|a, b)$ is assigned to the cluster. The product of all distributions thus assigned to a cluster then becomes the potential of the cluster (e.g., the product $P(c)P(d)$ for the cluster $\{b, c, d\}$ in T_0). A cluster that has no such assigned distributions is given a uniform potential (e.g., the cluster $\{a, b, c\}$

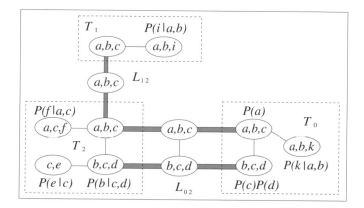

Figure 8.2: The LJF representation with initial potential assignment.

in T_1). Each separator in each local JT is given a uniform potential as well as each linkage in each linkage tree.

Once these three types of potentials (for clusters and separators of local JTs and linkages of linkage trees) are assigned, all other potentials are defined accordingly. These include the following potentials:

- A potential for each local JT.
- A potential for each separator in each linkage tree.
- A potential for each linkage tree.
- A joint system potential (JSP) for the entire LJF representation.

Although a single linkage tree L_{02} is drawn in Figure 8.2 between local JTs T_0 and T_2, in fact a separate linkage tree is obtained by agent A_0 from T_0 and another by A_2 from T_2. This is necessary because the two agents are in general remotely located and each needs a linkage tree to maintain its potential over the agent interface. Section 8.3 considers the consequences when the two linkage trees are computed by the two agents independently. The analysis concludes that, although in general the two linkage trees will be different, the difference is constrained and does not affect the result of belief updating by message passing.

The primitive action for message passing between two adjacent agents is the transmission of a potential over a single linkage. For example, the transmission of a potential $B(a, b, c)$ from A_0 to A_2 over the linkage $\{a, b, c\}$ in L_{02} is a primitive action (see Figure 8.2). Section 8.4 defines such a message.

Section 8.5 sets forth the actions involved in message passing between two adjacent agents over their linkage trees (recall that there are two of them – one for each agent). This process entails the transmission of potentials over linkages between agents followed by internal message passing in the receiving agent. For example, when A_0 passes its belief to A_2 over their interface, it sends a potential over the linkage $\{a, b, c\}$ and another potential over the linkage $\{b, c, d\}$. When the two potentials are received, A_2 combines the potential over the linkage $\{a, b, c\}$ with the potential in its linkage host (i.e., the cluster $\{a, b, c\}$) and combines the potential over the linkage $\{b, c, d\}$ with the potential in its linkage host. Afterwards, A_2 performs internal message passing over T_2, as discussed in Chapter 5.

Section 8.6 discusses how all agents in the system communicate through message passing over linkage trees. Figure 8.3 illustrates the process when the communication is activated by agent A_0. The black arrows indicate the sequence in which agents are activated: A_0 first, then A_2, and then A_1. When A_1 is activated, it passes the message to A_2 over their linkage trees. After local processing, as discussed above, A_2 passes the message to A_0 over their linkage trees. The white arrows indicate the sequence of the message passing. The first round of message passing is

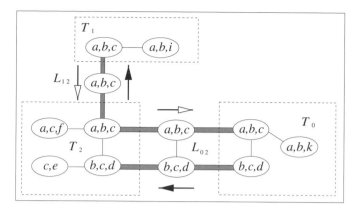

Figure 8.3: Communication among agents.

then completed, and the second round starts. After local processing, A_0 passes the message back to A_2, and A_2 passes the message to A_1 (following the black arrows). Section 8.6 shows that, after these two rounds of interagent message passing, all agents' beliefs are updated correctly.

The computational complexity of these communications is analyzed in Section 8.8. The conclusion is that as long as the MSBN structure is sparse, the multiagent communication operation is efficient.

To illustrate how multiple agents can use the MSBN–LJF representation and the communications algorithms presented in practice, Section 8.7 demonstrates troubleshooting a multi-component digital system by five cooperating agents. Two devices in the system are broken. By limited local observations and two communications, the agents are able to locate the two faulty devices with high certainty.

Section 8.9 discusses the issue of regional communication when not all agents in the multiagent system are involved in the communications activity. Regional communication is sometimes desirable because it is more efficient than full-scale communication, although agents that participate in it can only benefit from the knowledge of each other but not the agents outside the region. Regional communication is sometimes necessary when the computer network that physically connects agents fails. When such a failure occurs, regional communication allows agents isolated into each region to continue cooperation as a group; hence, the performance of the system degrades gracefully.

Section 8.10 examines the possibility of extending two alternative methods for belief updating in BNs, loop cutset conditioning and forward sampling, for belief updating in MSBNs. The analysis concludes that, although both extensions are possible in theory, they require many more messages to be transmitted among agents and much more complex coordination of agent activities.

8.2 Potentials in a Linked Junction Forest

To perform inference in an MSBN using the alternative dependence structure, an LJF, the JPD of the MSBN, needs to be converted into the belief over the LJF. As in the case of potential assignment for a JT in the single-agent paradigm (Section 5.3), we seek a concise and localized belief representation by exploring the independence (Theorem 7.15) encoded in the LJF. This involves assigning potentials to the clusters and separators of local JTs and linkage trees and assembling them into a joint system potential (JSP).

8.2.1 Defining Potentials

The following potentials are *assigned*, and all other potentials are *defined* (or *derived*) in terms of these potentials.

- Each cluster Q in each local JT is assigned a potential $B_Q(Q)$.
- Each separator S in each local JT is assigned a potential $B_S(S)$.
- Each linkage Q in each linkage tree is assigned a potential $B_Q(Q)$.

The initial assignment of these potentials is described in the next subsection. Here we define the other potentials first:

The potential of a JT T_i over the subdomain V_i is defined as

$$B_{T_i}(V_i) = \left[\prod_j B_{Q_j}(Q_j) \right] \Big/ \left[\prod_k B_{S_k}(S_k) \right],$$

where j is over the indices of all clusters in T_i and k is over the indices of all separators. The semantics of $B_{T_i}(V_i)$ have been established in Theorem 5.10.

For each separator S in each linkage tree, its potential is defined as

$$B_S(S) = \sum_{Q \backslash S} B_Q(Q),$$

where Q is any one of the two linkages with its separator S. Note that the definition refers to one of the two linkages without preference. As will be shown in Sections 8.3 and 8.5, this causes no problem at all. The potential of a linkage tree $L_{i,j}$ over a d-sepset $I_{i,j}$ is then defined as

$$B_{L_{i,j}}(I_{i,j}) = \left[\prod_m B_{Q_m}(Q_m) \right] \Big/ \left[\prod_k B_{S_k}(S_k) \right],$$

where m is over the indices of all linkages in $L_{i,j}$ and k is over the indices of all separators. In Section 8.2.3, we discuss the semantics of $B_{L_{i,j}}(I_{i,j})$.

The *joint system potential* (JSP) over the universe V is associated with the LJF F and is defined as

$$B_F(V) = \left[\prod_i B_{T_i}(V_i)\right] \Big/ \left[\prod_k B_{L_k}(I_k)\right],$$

where i is over the indices of all JTs in F and k is over the indices of linkage tree – one for each hyperlink in the hypertree. The semantics of $B_F(V)$ will be discussed in Section 8.2.3.

With the potentials attached to the JT and linkage trees of each local dependence structure (V_i, G_i, T_i, L_i), we refer to the resultant tuple $(V_i, G_i, T_i, B_{T_i}, L_i, B_{L_i})$ as the *local JT representation*. It is the compiled subdomain representation of the corresponding agent A_i. Note that B_{L_i} is a collection of linkage tree potentials, one for each linkage tree of the local structure. We refer to $\mathcal{F} = (F, B_F(V))$ as a *linked junction forest representation* of the corresponding MSBN.

8.2.2 Initial Potential Assignment

As described in the previous section, the potential for each cluster and each separator in each local JT and the potential for each linkage in each linkage tree are assigned (versus derived).

The potential assignment in each JT is the same as in the single-agent paradigm: Let T_i be a JT obtained from the subnet $S_i = (V_i, G_i, P_i)$. For each cluster Q and each separator S in T_i, assign a uniform potential. For each node v in G_i, find a cluster Q in T_i such that $fmly(v) \subseteq Q$ and break ties arbitrarily. Update $B(Q)$ to the product $B(Q) * P(v|\pi(v))$. Note that the assignment ensures

$$B_{T_i}(V_i) = P_i(V_i).$$

For each linkage in each linkage tree, assign a uniform potential.

Now consider the JSP

$$B_F(V) = \left[\prod_i B_{T_i}(V_i)\right] \Big/ \left[\prod_k B_{L_k}(I_k)\right].$$

Because each linkage is assigned a uniform potential, the denominator $\prod_k B_{L_k}(I_k)$ contributes no nontrivial information. Hence,

$$B_F(V) = \prod_i B_{T_i}(V_i).$$

Because each separator of each local JT is assigned a uniform potential, we have

$$B_F(V) = \prod_i B_{T_i}(V_i) = \prod_i P_i(V_i) = P(V). \qquad (8.1)$$

That is, the JSP is identical to the JPD of the MSBN.

Recall from Chapter 5 that supportiveness is an important property to ensure the correct belief updating by message passing. This is also true for belief updating in an LJF. That is, the initial potential assignment must ensure that each separator in each JT and each linkage tree is supportive (readers are encouraged to verify this). Each act of message passing must also maintain the supportiveness. Indeed, the inference algorithms presented later in this chapter do maintain the supportiveness. Because the analysis is straightforward, we omit it from our presentation.

8.2.3 Consistence of Potentials

Message passing achieves belief updating by bringing adjacent belief units into consistence. We define different levels of consistence in an LJF representation.

Consider a local JT T_i. If a pair of adjacent clusters Q and C and their separator S satisfy

$$\sum_{Q \setminus S} B_Q(Q) = \text{const}_1 * B_S(S) = \text{const}_2 * \sum_{C \setminus S} B_C(C),$$

then Q and C are said to be *consistent*. If every pair of adjacent clusters in T_i is consistent, then T_i is *locally consistent*. Note that this notion is paralleled in the single-agent paradigm. There, the local consistence ensures global consistence because of the JT structure. Here, we will use *global consistence* to describe the consistence at the level of the LJF and will not use the notion at the local JT level any more.

Because a linkage tree is a JT, we apply the notion of local consistence to a linkage tree in the same way as to a local JT.

Using the local consistence concept, the following proposition establishes the semantics of the linkage tree potential. That is, when a local JT is locally consistent, the potential of its linkage tree over a given d-sepset is the marginal of the JT potential. In other words, it is a correct e-message over the d-sepset.

Proposition 8.1 *Let T_i over V_i be a local JT in an LJF representation that is an I-map, and let T_i be locally consistent. Let $L_{i,j}$ be a linkage tree of T_i over the d-sepset $I_{i,j}$. For each linkage Q in $L_{i,j}$ with its host C in T_i, assign $B_Q(Q) =$*

$\sum_{C\setminus Q} B_C(C)$. *Then*

$$B_{L_{i,j}}(I_{i,j}) = \sum_{V_i\setminus I_{i,j}} B_{T_i}(V_i).$$

Proof: Because the LJF is an I-map, T_i is an I-map over V_i. Without losing generality, we assume that $B_{T_i}(V_i) = \text{const} * P(V_i)$. Then, by Theorem 5.16, for each cluster X in T_i, $B_X(X) = \text{const}_1 * P(X)$, and for each separator Z, $B_Z(Z) = \text{const}_2 * P(Z)$. Hence, according to the assignment of linkage potentials, for each linkage Q, we have $B_Q(Q) = \text{const}_3 * P(Q)$. According to the definition of separator potentials in a linkage tree, for each separator R in $L_{i,j}$ we have $B_R(R) = \text{const}_4 * P(R)$.

From Proposition 7.5, $L_{i,j}$ is a JT. From Proposition 7.6, $L_{i,j}$ is an I-map over $I_{i,j}$. From Theorem 5.10, the result follows. □

Next, we consider a local JT T_i and a linkage tree $L_{i,j}$ of T_i. If a linkage C in $L_{i,j}$ and its linkage host Q in T_i satisfy

$$\sum_{Q\setminus C} B_Q(Q) = \text{const} * B_C(C),$$

then C and Q are said to be *consistent*.

When T_i is locally consistent and each linkage in $L_{i,j}$ is consistent with its host in T_i, the local JT T_i and its linkage tree $L_{i,j}$ are said to be *consistent*. Recall that the separator potential in a linkage tree is defined by marginalizing the potential of its adjacent cluster. Hence, whenever T_i and $I_{i,j}$ are consistent, $L_{i,j}$ is also locally consistent. From Proposition 8.1, when T_i and $L_{i,j}$ are consistent, $L_{i,j}$ carries all the relevant information that agent A_i has relative to agent A_j.

Finally, consider an LJF representation $\mathcal{F}$. If every local JT is consistent and every linkage tree is consistent with its local JT, then $\mathcal{F}$ is said to be *globally consistent*. When $\mathcal{F}$ reaches global consistence, every agent will have received all relevant information that other agents have to offer. Further communication would have no effect on each agent's belief. Using the notion of global consistence, we establish the semantics of JSP in the remainder of this section.

Theorem 8.2 *Let a hypertree MSDAG $G = \sqcup_i G_i$ be an I-map of a total universe V and F be an LJF of G. If for each cluster Q in each local JT of F, $B_Q(Q) = \text{const}_1 * P(Q)$ and the LJF representation $(F, B_F(V))$ is globally consistent, then*

$$P(V) = \text{const} * B_F(V).$$

Proof: Recall that

$$B_F(V) = \left[\prod_i B_{T_i}(V_i) \right] \bigg/ \left[\prod_k B_{L_k}(I_k) \right].$$

By Theorem 7.15, F is an I-map in the sense that the hypertree is an I-map, each local JT is an I-map, and each linkage tree is also an I-map. Because each local JT is an I-map, we have

$$P(V_i) = \text{const}_2 * B_{T_i}(V_i),$$

according to Theorem 5.10. Inasmuch as each T_i is locally consistent and is consistent with each of its linkage trees $L_{i,j}$, by Proposition 8.1 we have

$$P(I_{i,j}) = B_{L_{i,j}}(I_{i,j}).$$

Because the hypertree is an I-map, we have

$$P(V) = \text{const} * \left[\prod_i P(V_i) \right] \bigg/ \left[\prod_k P(I_k) \right]. \qquad \square$$

Recall Basic Assumption 6.5. It requires the uncertain knowledge representation of a multiagent system to respect the belief of each agent within its subdomain and to supplement each agent's knowledge with others' outside the subdomain. Theorem 8.2 assures us that when an MSBN is compiled into an LJF representation, Basic Assumption 6.5 is kept in this alternative representation.

8.3 Linkage Trees over the Same d-sepset

Recall that each agent is associated with a local JT and a set of linkage trees – one for each d-sepset. Because two adjacent agents have a shared d-sepset $I_{i,j}$, each of them has a linkage tree over the d-sepset. Given that each agent computes its linkage tree over $I_{i,j}$ independently from its local JT, the two linkage trees are potentially different. We analyze the potential difference, which has implications for e-message passing operations during agent communication. We consider the following possible difference between the two linkage trees:

1. The linkages of the two trees may be different.
2. The separators of the two trees may be different.
3. When the clusters and the separators are identical, the topology of the two trees may still be different.
4. When the clusters and the separators are identical, the potentials of the two trees may still be different.

First, we consider the linkages. According to Theorem 7.12, the adjacent local graphs G_i and G_j are graph-consistent. Hence, the subgraphs of G_i and G_j spanned by the d-sepset are identical. Each linkage is a clique in the subgraph. Hence, the linkages in the two trees must be identical.

Second, we consider the separators. According to Proposition 7.5, each linkage tree is a junction tree with the linkages as clusters. We have shown in the preceding paragraph that the clusters of the two linkage trees are identical. The following proposition shows that all junction trees of the same set of clusters have the same set of separators as well.

Proposition 8.3 *Let G be a connected chordal graph, and H and T be two distinct corresponding JTs of G. Then H and T have the identical set of separators.*

Proof: Because the clusters of H are cliques of G and so are clusters of T, H and T have the same set of clusters. We prove the proposition by induction on the number k of clusters of H. The statement trivially holds if $k = 2$. Assume that it holds when $k = n$. We consider H with $n + 1$ clusters.

Let C be any terminal cluster of H with the adjacent cluster C^* and the separator S, as in Figure 8.4(a). Let H' be the JT resultant from removing the cluster C (and its separator) from H.

If C is terminal in T, let T' be the JT resultant from removing C from T.

Consider the case in which C is not terminal in T, as in Figure 8.4(b). Because T is a JT, every separator on the path between C and C^* must be equal to S. Hence, at least one separator incident to C in T is S. Denote the corresponding adjacent cluster of C as Q. Remove C and its separator S with Q, and let other adjacent clusters of C be adjacent to Q, as shown in (c). The resultant is still a JT. Denote this JT as T'.

Clearly H' and T' have n clusters each. By the inductive assumption, they have the identical set of clusters. Because the only separator removed from H and T is S, the result follows. $\square$

Third, we recognize that the topology of the two linkage trees may be different even when both have the same clusters and separators. This means that the same

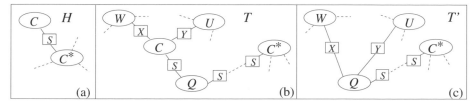

Figure 8.4: Illustration of proof for Proposition 8.3.

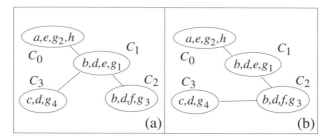

Figure 8.5: Two JTs with identical clusters and separators but different topologies.

cluster in different linkage trees may have different separators. Figure 8.5 shows such an example, where C_1 has three separators in (a) but two in (b). On the other hand, the potential of a linkage tree is defined in terms of potentials over its linkages and separators. Hence, as long as the linkages and separators are identical, the topological difference has no affect on the linkage tree potential.

Finally, we consider the potential value. Clearly, because each linkage tree is maintained by a separate agent, it is possible from time to time that the two linkage tree potentials are inconsistent. Recall that the JSP of an LJF F is defined as

$$B_F(V) = \left[\prod_i B_{T_i}(V_i) \right] \Big/ \left[\prod_k B_{L_k}(I_k) \right].$$

The denominator involves a linkage tree potential for every d-sepset. If the two linkage trees are inconsistent, the JSP would not be uniquely defined. We have seen that the initial assignment does ensure that each pair of linkage trees over the same d-sepset is consistent (by assigning them uniform potentials). As we will see, during communication the consistence will be maintained.

8.4 Extended Linkage Potential

In this section, we consider an extended potential to be associated with each linkage. Recall that the potential of a linkage tree $L_{i,j}$ over a d-sepset $I_{i,j}$ is defined as

$$B_{L_{i,j}}(I_{i,j}) = \left[\prod_m B_{Q_m}(Q_m) \right] \Big/ \left[\prod_k B_{S_k}(S_k) \right],$$

where m is over the indices of all linkages in $L_{i,j}$ and k is over the indices of all separators. When the linkage tree is locally consistent, each separator potential carries redundant information because it is simply a marginal of some cluster potential. Their appearance as denominators in the preceding equation can be interpreted as a "removal" of the redundance from the numerator, the product of cluster potentials.

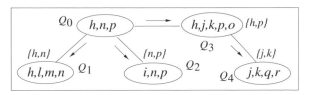

Figure 8.6: Defining linkage peers for each linkage.

Instead of removing redundance after the product, the extended potential explores the idea of removing redundance before the product.

First, we introduce the notion of *peer separator* of a linkage to signify what redundant information needs to be removed and from where.

Definition 8.4 *Let $L_{i,j}$ be a linkage tree of a local JT. Convert $L_{i,j}$ into a rooted tree by selecting a linkage Q arbitrarily as the root and direct links away from it. For each linkage $Q' \neq Q$, assign the separator with its parent linkage as the **peer separator** of Q'.*

For example, the linkage tree in Figure 8.6 has five linkages. If the linkage Q_0 is chosen as the root, then it has no peer assigned to it. The peer separator of each other linkage is labeled beside the linkage. We now define the extended linkage potential.

Definition 8.5 *Let $L_{i,j}$ be a linkage tree with linkage potentials, separator potentials, and linkage peers defined. For each linkage Q with peer R, its **extended linkage potential** is*

$$B^*_Q(Q) = B_Q(Q)/B_R(R).$$

For the cluster Q without peer, define

$$B^*_Q(Q) = B_Q(Q).$$

Consider the linkage tree in Figure 8.6. The extended potential for linkage Q_1 is

$$B^*_{Q_1}(h, l, m, n) = B_{Q_1}(h, l, m, n)/B_{\{h,n\}}(h, n),$$

and the extended potential for linkage Q_0 is $B^*_{Q_0}(h, n, p) = B_{Q_0}(h, n, p)$.

The semantics of extended linkage belief are shown in Proposition 8.6. The proof is trivial.

Proposition 8.6 *Let $L_{i,j}$ be a linkage tree. Then the linkage tree potential $B_{L_{i,j}}(I_{i,j})$ can be expressed in terms of extended linkage potentials as*

$$B_{L_{i,j}}(I_{i,j}) = \prod_Q B_Q^*(Q),$$

where each Q is a linkage in $L_{i,j}$.

In the following sections, the extended linkage potential $B_Q^*(Q)$ will be used as the actual e-message transmitted between agents, whereas the linkage potential $B_Q(Q)$ and separator potential $B_S(S)$ will only be used as conceptual objects in our analysis. In our presentation, whenever either $B_Q(Q)$ or $B_Q^*(Q)$ is updated, it is assumed that both are updated *simultaneously*.

8.5 E-message Passing between Agents

In this section, we describe how agents in a system organized as an LJF representation achieve consistence by message passing. First, we introduce **AbsorbThrough-Linkage**, whose effect is to propagate belief from one agent to an adjacent agent through a single linkage.

Algorithm 8.1 (AbsorbThroughLinkage) *Let A_i and A_j be adjacent agents. Agent A_i is associated with the local JT T_i and linkage tree L_i, and A_j with T_j and L_j. Let Q_i be a linkage in L_i, C_i be the linkage host of Q_i in T_i, and Q_j be the corresponding linkage in L_j.*

*When **AbsorbThroughLinkage** is called on A_i for C_i to absorb through Q_i, the following occurs:*

1. *Agent A_i requests transmission of $B_{Q_j}^*(Q_j)$ from A_j.*
2. *Upon receipt, A_i updates its host potential $B_{C_i}'(C_i) = B_{C_i}(C_i) * B_{Q_j}^*(Q_j)/B_{Q_i}^*(Q_i)$.*
3. *Agent A_i updates its linkage potential $B_{Q_i}^{*'}(Q_i) = B_{Q_j}^*(Q_j)$.*

The following **UpdateBelief** algorithm propagates belief from an agent to an adjacent agent through multiple linkages (a hyperlink) between them. **UpdateBelief** is analogous to **Absorption** in Section 5.4 for message passing in a single agent JT. However, here, the message originates from a local JT and is targeted at another local JT. The channel is a d-sepset and (a hyperlink) in the form of multiple linkages.

Algorithm 8.2 (UpdateBelief) *Let A_i and A_j be adjacent agents. Agent A_i is associated with local JT T_i and linkage tree L_i, and A_j with T_j and L_j. When*

UpdateBelief is called on A_i relative to A_j, the following occur:

1. Upon request from A_i, for each linkage Q_j with host C_j in L_j, A_j assigns $B_{Q_j}(Q_j) = \sum_{C_j \backslash Q_j} B_{C_j}(C_j)$.
2. For each linkage Q_i with host C_i in L_i, A_i calls **AbsorbThroughLinkage** on itself for C_i to absorb through Q_i.
3. Agent A_i performs **UnifyBelief** (Algorithm 5.4) at T_i.

From Propositions 8.1 and 8.6, the messages passed through all linkages in step 2 collectively define an e-message, the belief of agent A_j over the d-sepset with A_i. Hence, **UpdateBelief** essentially performs e-message passing between a pair of agents. Its effects are shown formally in the following proposition.

Proposition 8.7 *Let T_i and T_j be adjacent local JTs, and L_i and L_j be the corresponding linkage trees. Let T_j be locally consistent. After **UpdateBelief** is performed in A_i relative to A_j, the following hold:*

1. *The JT T_i is locally consistent.*
2. *Linkage trees L_i and L_j are both consistent with T_j.*
3. *If L_i and L_j were consistent before **UpdateBelief**, the JSP of the LJF is invariant after **UpdateBelief**.*
4. *If T_i was locally consistent and was consistent with L_i before **UpdateBelief**, T_i and L_i are also consistent after **UpdateBelief**.*

Proof:
1. This holds due to **UnifyBelief** at the end of **UpdateBelief**.
2. Linkage tree L_j is consistent with T_j due to the first step of **UpdateBelief**. Linkage trees L_i and L_j are consistent because of the last step of **AbsorbThroughLinkage**.
3. Because L_i and L_j were consistent, the JSP is uniquely defined irrespective of the potential of the linkage tree referenced:

$$B_F(V) = \left[\prod_m B_{T_m}(V_m) \right] \Big/ \left[\prod_k B_{L_k}(I_k) \right].$$

After **UpdateBelief**, the potentials of L_i, L_j, and T_i are updated, and as shown in (2), L_i and L_j are consistent. Hence, only terms $B_{T_i}(V_i)$ and $B_{L_i}(I_i)$ in the preceding expression are modified. The updated $B_{L_i}(I_i)$ can be equivalently represented as

$$B'_{L_i}(I_i) = B_{L_i}(I_i) * \left[B'_{L_i}(I_i) \big/ B_{L_i}(I_i) \right].$$

The updated potential for T_i is

$$B'_{T_i}(V_i) = B_{T_i}(V_i) * \left[B'_{L_i}(I_i) \big/ B_{L_i}(I_i) \right].$$

Therefore, after **UpdateBelief**, the JSP is invariant.

4. For simplicity, we assume that all constants involved have a value of 1. We need to show

$$\sum_{V_i \setminus I_i} B'_{T_i}(V_i) = B'_{L_i}(I_i).$$

This is true because

$$\sum_{V_i \setminus I_i} B'_{T_i}(V_i) = \left(B'_{L_i}(I_i)/B_{L_i}(I_i)\right) * \sum_{V_i \setminus I_i} B_{T_i}(V_i) = \left(B'_{L_i}(I_i)/B_{L_i}(I_i)\right) * B_{L_i}(I_i) = B'_{L_i}(I_i),$$

where the first equation is derived from Theorem 5.8 and the second one from the local consistency of T_i and consistency between T_i and L_i before **UpdateBelief**.   □

Recall from Section 8.3 that because the two linkage trees over a d-sepset are computed independently by two corresponding agents, they may differ in topology. In addition, from Section 8.4, the peer of each linkage is also computed by the corresponding agent independently. Hence, the two agents may define the extended linkage potential differently for each linkage (owing to difference in topology or in the choice of root when defining the linkage peers). The preceding analysis shows that such differences are immaterial. This observation is useful, for otherwise the two agents would have to coordinate the topology of their linkage trees as well as the choice of root when defining the linkage peers.

8.6 Multiagent Communication

Through a sequence of e-message passing among agents organized into an LJF, the global consistency in the multiagent system can be achieved. We present the agent communication algorithms in this section.

The following **CollectBelief** algorithm recursively propagates belief inwards (from terminal agents towards an initiating agent) on the hypertree of an LJF. Just as **UpdateBelief** is analogous to **Absorption** at a higher abstraction level, **CollectBelief** is analogous to **CollectEvidence** in the single-agent JT propagation but is at the hypertree level.

Algorithm 8.3 (CollectBelief) *Let A_0 be an agent with local JT T_0. A **caller** is either an adjacent agent A_c or the system coordinator. Denote additional adjacent agents of A_0 by $A_1, \ldots, A_m$ if any. When the caller calls on A_0 to **CollectBelief**, it does the following:*

1. *If A_0 has no adjacent agent except caller, it performs **UnifyBelief** at T_0 and returns.*
2. *Otherwise, for each agent A_i $(i = 1, \ldots, m)$, A_0 calls **CollectBelief** on A_i. After A_i finishes, A_0 calls on itself to **UpdateBelief** relative to A_i.*

The following **DistributeBelief** algorithm recursively propagates belief outwards (from an initiating agent towards terminal agents) on the hypertree of an LJF. **DistributeBelief** is analogous to **DistributeEvidence** in the single-agent JT propagation but is at the hypertree level.

Algorithm 8.4 (DistributeBelief) *Let A_0 be an agent with local JT T_0. A **caller** is either an adjacent agent A_c or the system coordinator. Denote additional adjacent agents of A_0 by $A_1, \ldots, A_m$, if any. When the caller calls on A_0 to **DistributeBelief**, it does the following:*

1. *If caller is an adjacent agent, A_0 calls **UpdateBelief** on itself relative to caller.*
2. *For each agent A_i ($i = 1, \ldots, m$), A_0 calls **DistributeBelief** on A_i.*

The following **CommunicateBelief** algorithm combines **CollectBelief** and **DistributeBelief. CommunicateBelief** is analogous to **UnifyBelief** in the single-agent JT propagation but is at the hypertree level.

Algorithm 8.5 (CommunicateBelief) *An LJF representation is populated by multiple agents with one at each hypernode. The system coordinator does the following:*

*Choose an agent A_* arbitrarily. Call **CollectBelief** on A_* followed by a call of **DistributeBelief** on A_*.*

CommunicateBelief brings an LJF representation into global consistency, as shown in Theorem 8.9 below. Our proof uses the following notation (see Figure 8.7 for illustration). Let $A_i \neq A_*$ be an agent on the hypertree. Let A_j be the adjacent agent of A_i on the path between A_i and A_*. Denote their local JTs by T_i, T_j, and T_*, respectively. Denote the linkage tree of A_i relative to A_j by L_i and that of A_j relative to A_i by L_j.

Lemma 8.8 *After **CollectBelief** is called on A_*, T_i is locally consistent and is consistent with L_i and L_j.*

Proof: If A_i is a terminal agent on the hypertree, when **CollectBelief** is called on A_i, it performs **UnifyBelief** at T_i. Hence, T_i is locally consistent by Theorem 5.14.

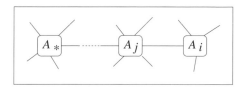

Figure 8.7: Illustration of notation for proof of Theorem 8.9.

If A_i is an internal agent on the hypertree, it performs **UpdateBelief** relative to each adjacent agent except the caller A_j. Because the last step in each **UpdateBelief** executes **UnifyBelief**, it renders T_i locally consistent.

When A_j is called to perform **CollectBelief**, it performs an **UpdateBelief** relative to A_i. Before the **UpdateBelief**, T_i is locally consistent, as shown. Hence, **UpdateBelief** renders L_i and L_j consistent with T_i by Proposition 8.7 (2).   □

Using the lemma, we now prove Theorem 8.9, which establishes the correctness of **CommunicateBelief**.

Theorem 8.9 *After **CommunicateBelief** is applied to an LJF representation* $(F, B_F(V))$, *it is globally consistent and* $B_F(V)$ *is invariant.*

Proof: To prove global consistence, we show that for an arbitrary pair of adjacent agents A_i and A_j (Figure 8.7), T_i and T_j are locally consistent, and each is consistent with both L_i and L_j.

By Lemma 8.8, after **CollectBelief** is called on A_*, each local JT T_i is locally consistent, and each pair of corresponding linkage trees L_i and L_j are consistent with T_i.

Consider the state after **DistributeBelief** is called on A_*. When A_j is called to perform **DistributeBelief**, it performs an **UpdateBelief** relative to the caller, which renders T_j locally consistent by Proposition 8.7 (1).

After A_i is called by A_j to perform **DistributeBelief**, T_i is locally consistent due to the **UpdateBelief** relative to A_j, and T_j is consistent with L_i and L_j by Proposition 8.7 (2). Because L_i was consistent with T_i, as shown above for **CollectBelief**, L_i is consistent with T_i as well owing to Proposition 8.7 (4).

Next, consider the invariance of $B_F(V)$. **CommunicateBelief** consists of a set of **UpdateBelief**. By Proposition 8.7 (3), $B_F(V)$ is invariant.   □

The following theorem shows that when an LJF representation is globally consistent, each local JT contains the correct marginal over the corresponding subdomain.

Theorem 8.10 *Let* $\mathcal{F} = (F, B_F(V))$ *be an LJF representation, where F is an I-map over V. Let* T_i *be a local JT over* V_i *in* $\mathcal{F}$. *If* $\mathcal{F}$ *is globally consistent, then*

$$B_{T_i}(V_i) = \text{const} * \sum_{V \setminus V_i} B_F(V).$$

Proof: For simplicity, we assume that all the constants involved have the value 1. The JSP is

$$B_F(V) = \left[\prod_{T_m \text{ in } F} B_{T_m}(V_m) \right] \bigg/ \left[\prod_{L_k \text{ in } F} B_{L_k}(I_k) \right].$$

Let T_j be a terminal hypernode on the hypertree of $\mathcal{F}$ and I_j be the corresponding d-sepset. Let $\mathcal{F}'$ be the LJF representation obtained by removing T_j (and I_j) from $\mathcal{F}$ and $V' = V \setminus (V_j \setminus I_j)$. Because F is an I-map, we have

$$B_{F'}(V') = \sum_{V_j \setminus I_j} \left[\prod_{T_m \text{ in } F} B_{T_m}(V_m) \right] \bigg/ \left[\prod_{L_k \text{ in } F} B_{L_k}(I_k) \right]$$

$$= \left\{ \left[\prod_{T_m \text{ in } F'} B_{T_m}(V_m) \right] \bigg/ \left[\prod_{L_k \text{ in } F'} B_{L_k}(I_k) \right] \right\} * \sum_{V_j \setminus I_j} \left[B_{T_j}(V_j) / B_{L_j}(I_j) \right]$$

$$= \left[\prod_{T_m \text{ in } F'} B_{T_m}(V_m) \right] \bigg/ \left[\prod_{L_k \text{ in } F'} B_{L_k}(I_k) \right],$$

where the first equation obtains $B_{F'}(V')$ from $B_F(V)$ by marginalizing out the variables in $V_i \setminus V'$, the second equation applies Theorem 5.8, and the third equation results from consistence between T_j and L_j; F' is still a hypertree and an I-map over V'.

By recursively applying the marginalization to a terminal hypernode of the resultant hypertree, eventually we obtain

$$B_{T_i}(V_i) = \sum_{V \setminus V_i} B_F(V).$$

$\square$

Recall from Eq. (8.1) in Section 8.2.2 that, after the initial potential assignment for an LJF representation obtained from an MSBN, we have $B_F(V) = P(V)$. From Theorem 5.16, when a JT is locally consistent, the potential associated with each cluster is the correct marginal of the potential over the JT. Combining these with Theorem 8.10, we have the following corollary, which says that when an LJF is globally consistent, the potential associated with each cluster in each local JT is the correct marginal of the JPD of the deriving MSBN.

Corollary 8.11 *Let $\mathcal{F}$ be an LJF representation derived from an MSBN over V. If $\mathcal{F}$ is globally consistent, then for each cluster C in each local JT,*

$$B_C(C) = \text{const} * \sum_{V \setminus C} P(V),$$

where $P(V)$ is the JPD of the MSBN.

Based on Corollary 8.11, the prior probability of any variable in an MSBN can be obtained by marginalization of any cluster that contains the variable in any local JT of the LJF. Next, we consider how to obtain posteriors after agents make observations.

Recall from Section 5.6 that, in the case of a single-agent JT, if the agent observes the values of a subset X of variables, **EnterEvidence** can be performed to update the potential of the JT into $B_T(V) = \text{const} * P(V|X = \mathbf{x})$. In a multiagent system, agents make observations in their subdomains independently. Let $X_i \subset V_i$ be the subset of variables observed by agent A_i. Let $\mathbf{x}$ be the configuration of $X = \cup_i X_i$ corresponding to the observation by all agents. Extending the discussion in Section 5.6, we have

$$P(V|\mathbf{x}) = B_F(V) * \prod_i \prod_{x \in X_i} obs(x),$$

which is equivalent to

$$\prod_i \left\{ \left[B_{T_i}(V_i) * \prod_{x \in X_i} obs(x) \right] \Big/ \left[\prod_k B_{L_k}(I_k) \right] \right\}.$$

This means that the posterior JPD over V can effectively be obtained simply by making each agent A_i perform **EnterEvidence** *locally* for its observation on X_i.

After each agent has finished, the LJF is *no longer* globally consistent. If another **CommunicateBelief** is activated in the LJF, it will again be globally consistent. Applying Theorems 8.10 and 5.16, we have the following results similar to Corollary 8.11 but applicable to posteriors.

Theorem 8.12 *Let $\mathcal{F}$ be a globally consistent LJF representation over V and $P(V)$ be the JPD. After all agents with local observations have performed **EnterEvidence** locally, a **CommunicateBelief** is performed in $\mathcal{F}$. Then for each cluster C in each local JT,*

$$B_C(C) = \text{const} * \sum_{V \setminus C} P(V|\mathbf{x}),$$

where the configuration $\mathbf{x}$ denotes all the observations entered.

Corollary 8.11 and Theorem 8.12 justify the following operations: After an LJF representation is converted from an MSBN and **CommunicateBelief** is performed, priors for any variables in any subdomain can be obtained locally by the corresponding agent. After several agents have entered their observations, if another **CommunicateBelief** is performed, posteriors can be obtained locally by each agent. This process can repeat any finite number of times. Between two consecutive executions of **CommunicateBelief**, if an agent has local observations, by performing **EnterEvidence** followed by **UnifyBelief**, it can obtain posteriors that are exact relative to all the observations accumulated in the entire system up to the last **CommunicateBelief** and relative to all the local observations by the agent.

8.7 Troubleshooting a Digital System

In this section, we demonstrate how an MSBN-based multiagent system functions in practice. We will use digital system monitoring as the problem domain. The physical components of the digital system are shown in Figures 6.6 through 6.10. Five agents populate the monitoring system. The virtual components are shown in Figures 6.14 through 6.17 and Figure 6.10. The subnet structures are shown in Figures 6.27 through 6.31. The hypertree is shown in Figure 6.21(a). The local junction trees are illustrated in Figures 7.29 through 7.33, and the linkage trees are presented in Figures 7.34 through 7.37.

In addition to the dependence structures, we assume the following representational parameters: Each logical gate is represented as a binary variable and is either normal or faulty. For simplicity, the space is denoted as {good, bad}. It is assumed that each gate has a 0.01 probability of being faulty. A faulty gate is modeled so that it may or may not produce the incorrect output. A faulty AND gate is assumed to output correctly 20% of the time. A faulty OR gate outputs correctly 70% of the time, and a faulty NOT gate outputs correctly 50% of the time. We consider a scenario in which the external inputs are as follows:

$$a_0 = 0, a_2 = 1, b_0 = 0, e_2 = 1, h_2 = 0, i_0 = 1, i_2 = 1,$$
$$k_0 = 0, k_1 = 0, l_2 = 0, o_0 = 0,$$
$$o_1 = 0, o_2 = 1, p_1 = 1, s_0 = 0, v_4 = 1, v_7 = 1, y_1 = 1,$$
$$y_2 = 1, z_1 = 0, z_5 = 0.$$

In addition, suppose that the gates d_1 and t_5 are faulty and produce incorrect outputs. The state of the total universe is then completely defined and is shown in Figure 8.8. The input and output of each gate are labeled in the figure. Because gates d_1 and t_5 produce incorrect outputs, the outputs of other gates downstream will also be affected. Each incorrect output is underlined in the figure.

Note that the view in Figure 8.8 is only for the convenience of the reader. This view is not available to any of the agents. We assume that the state of each gate is not observable by any agent. We also assume that observation of each input and output has a cost. Hence, observing all inputs and outputs is not an option. To make the situation more challenging, we assume that a direct input x_4 of the faulty gate t_5 and the direct output w_0 of the faulty gate d_1 are unobservable.

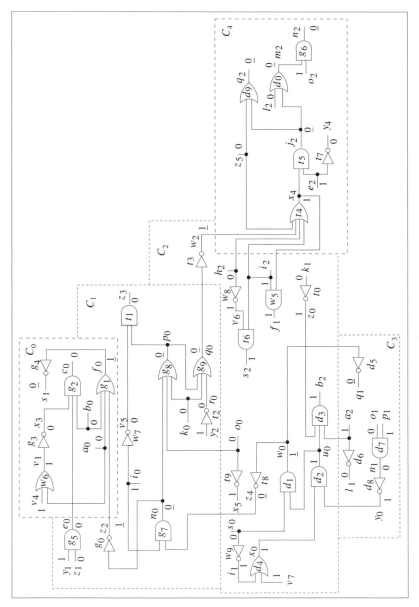

Figure 8.8: The inputs and outputs of all gates with incorrect outputs underlined.

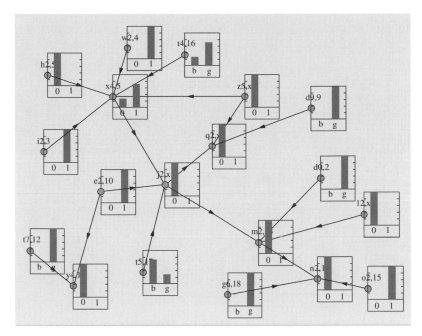

Figure 8.9: Agent A_4 cannot decide if t_4 or t_5 is abnormal.

First, we demonstrate the limitation of autonomous inference by individual agents without cooperation. Consider agent A_4. The belief of A_4 after all variables have been observed, except x_4 and the states of gates, is shown in Figure 8.9. The height of each histogram ranges from 0 to 1. The labels 0 and 1 represent logical values of the inputs and outputs of each gate. The labels b and g stand for the states of gates: bad and good.

Because x_4 is unobservable, although agent A_4 suspects that t_4 or t_5 might be faulty, it is quite uncertain (with belief 0.28 and 0.73 for each being faulty, as shown by the histogram beside each corresponding node). Hence, A_4 is unable to justify a replacement action with high certainty. We will see next that by cooperating with other agents, A_4 can do much better.

For the cooperative monitoring, suppose that initially each agent makes some local observations, as shown in Table 8.1.

Table 8.1: *Local observations by each agent*

A_0 :	a_0,	b_0,	s_1	
A_1 :	i_0,	k_0,	z_3	
A_2 :	i_2,	s_0,	b_2	
A_3 :	o_1,	l_1		
A_4 :	i_2,	e_2,	o_2,	n_2

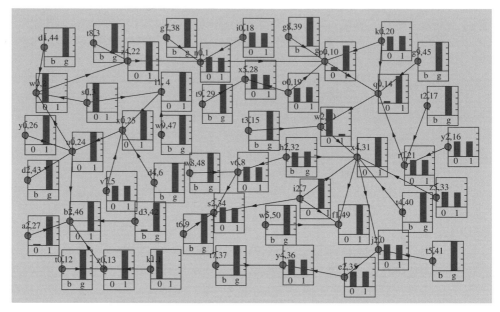

Figure 8.10: Agent A_2 sees nothing wrong before communication.

Owing to the limited amount of observations, most agents do not yet detect any abnormality. Figure 8.10 shows the belief of A_2. All gates are believed to be normal with high probability.

Only A_4 detects that something is wrong. Its probabilities for gates t_4, t_5, d_0, and g_6 to be faulty are 0.10, 0.25, 0.19, and 0.49, respectively, as shown in Figure 8.11. Agent A_4 then initiates a communication. Agent A_4 can achieve this by assuming

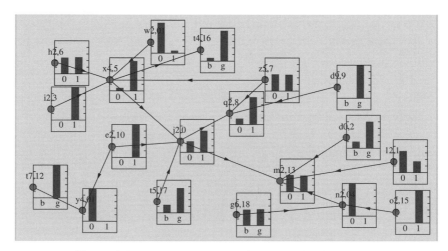

Figure 8.11: Agent A_4 detects that something is wrong.

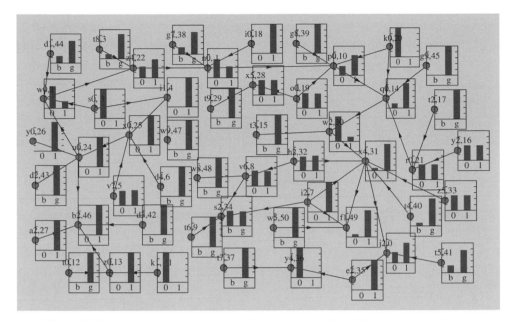

Figure 8.12: Agent A_2 suspects gates d_1, t_8, g_7, t_4, and t_5 after communication.

the role of the system coordinator and activating **CommunicateBelief** with A_* being itself.

After the communication, the deviation of circuit outputs from their expected values is detected by all agents. Agent A_0's belief on gates g_1 and g_4 being faulty are 0.09 and 0.15, respectively. Agent A_1's belief on gates g_0, g_1, and g_7 being faulty are 0.15, 0.09, and 0.24, respectively. Agent A_2 suspects abnormality in gates d_1, t_8, and g_7 as well as gates t_4 and t_5 that physically belong to the subdomain of A_4 (see Figure 8.12). Agent A_3's belief in gate d_1 being faulty is 0.24. Agent A_4's belief on gates t_4, t_5, g_6, and d_0 being faulty are 0.10, 0.25, 0.49, and 0.19, respectively. In summary, many gates are suspected but not conclusively.

To further reduce the uncertainty, suppose that each agent makes one more observation. Agent A_1 observes q_0, and its belief is shown in Figure 8.13. The agent has now ruled out gates g_1, g_0, and g_9 but still suspects g_7.

Figure 8.14 shows that A_3 is quite confident that d_1 is faulty (with belief 0.98) after making an additional observation on q_1.

After observing c_0, A_0's belief on gates g_1 and g_4 does not change much. After observing f_1, A_2 rules out t_4 with its belief on t_5's being faulty increased to 0.27. Its uncertainty about gates d_1, t_8, and g_7 is unchanged. Agent A_4 observes q_2 and rules out gates d_0 and g_6. It still suspects t_4 (with belief 0.27) and t_5 (with belief 0.72).

The unresolved uncertainty demands another communication. Suppose that A_1 initiates the communication this time. After that, agents A_0, A_1, and A_3 are certain

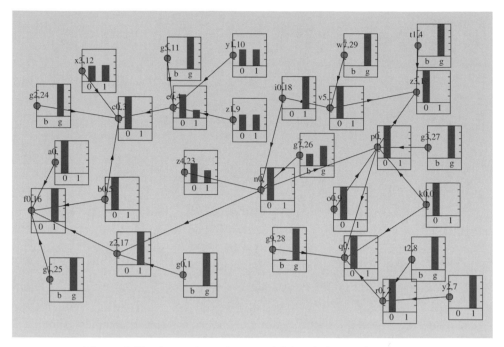

Figure 8.13: Agent A_1 makes an additional observation on q_0.

that all gates in their physical domains are normal. Agent A_2 is confident that only d_1 is abnormal. Agent A_4 is sure that gate t_5 is faulty and everything else is normal, as shown in Figure 8.15. On average, four observations have been made per agent. The cooperation allows agents to converge their belief to high certainty even though some outputs of the gates are unobservable.

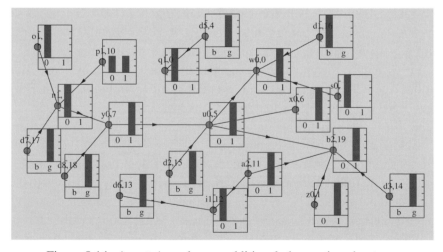

Figure 8.14: Agent A_3 makes an additional observation about q_1.

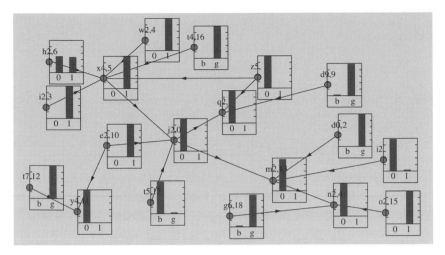

Figure 8.15: Agent A_4 knows that t_5 is faulty after the second communication.

8.8 Complexity of Multiagent Communication

We consider the time complexity of **CommunicateBelief** in a multiagent system organized into an LJF. We used the following parameters to characterize the LJF.

- n: the total number of agents in the system.
- m: the maximum number of clusters in a local JT.
- q: the cardinality of the largest cluster in local JTs.
- r: the maximum number of linkages in a linkage tree.

During **CommunicateBelief**, **UpdateBelief** is performed twice for each hyperlink in the hypertree, once during **CollectBelief**, and once during **DistributeBelief**. Hence, **UpdateBelief** is performed $2(n - 1)$ times because a tree of n nodes has $n - 1$ links.

UpdateBelief has three steps. The first step updates up to r local linkage potentials (one for each linkage) and has a complexity of $O(r\, 2^q)$. In the second step, **AbsorbThroughLinkage** is performed up to r times (one for each linkage). For each **AbsorbThroughLinkage**, an extended linkage potential is transmitted, and a linkage host potential is updated. Hence, this step has the complexity $O(r\, 2^q)$. In the last step of **UpdateBelief**, **UnifyBelief** is performed, which has the complexity $O(m\, 2^q)$ (see Section 5.5). The overall complexity of **UpdateBelief** is then

$$O((m + 2r)\, 2^q).$$

Combining the preceding analysis, we arrive at the conclusion that the complexity of **CommunicateBelief** is $O(n\,(m + 2r)\, 2^q)$. It is assumed that the shared variables between a pair of agents are a small subset of either subdomain involved (see Exercise 3). This implies $m \gg r$. Therefore, the complexity of **CommunicateBelief** can be simplified to

$$O(n \, m \, 2^q).$$

It is efficient as long as the cardinality q of the largest cluster is reasonably small.

8.9 Regional Multiagent Communication

Although **CommunicateBelief** can be performed effectively, when the problem domain is very large, the computation of **CommunicateBelief** can still be quite expensive. Each agent must pass i-messages locally during **UnifyBelief**. To pass e-messages among agents, **AbsorbThroughLinkage** must be performed, which involves transmission across some media, resulting in delay. Furthermore, the e-message passing and local i-message passing must be performed in a semiparallel fashion, as illustrated in Figure 8.16.

Consider the LJF in Figure 8.16. Suppose that **CommunicateBelief** is activated at A_0. During **CollectBelief**, the local computation at each agent and e-message passing among agents on the hyperchain $\langle A_0, A_1, A_2, A_3, A_4 \rangle$ occur strictly in sequence. That is, the local computation at A_4 is followed by e-message passing from A_4 to A_3, followed by local computation at A_3, followed by e-message passing from A_3 to A_2, and so on until eventually the local computation takes place at A_0. The overall **CollectBelief** is semiparallel in that the computation along the hyperchain $\langle A_1, A_2, A_3, A_4 \rangle$ occurs in parallel with, say, that along $\langle A_1, A_8, A_9 \rangle$. The similar semiparallel and semisequential computation occurs during **DistributeBelief** but in the reverse direction.

The semiparallel computation implies that the computational time of **CommunicateBelief** is lower bounded by the length of the longest hyperchain in the LJF. In general, a given agent's belief about its subdomain is more sensitive to the probabilistic information contained in agents close to it on the hypertree than those

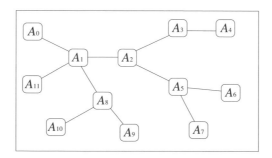

Figure 8.16: Illustration of semiparallel computation during **CommunicateBelief**.

located far away (see Exercise 4 for exceptional cases). Therefore, it makes sense for an agent to exchange belief with only nearby agents on the hypertree before an urgent decision without activating the full scale of **CommunicateBelief**. That is, a regional communication is sometimes desirable, for it is more efficient.

When agents are distributed, their physical connection through a computer network may fail from time to time. When that happens, it is impossible to conduct a full-scale **CommunicateBelief**. However, each agent can still benefit from the information contained in other agents that are still physically connected to it. In such a case, a regional **CommunicateBelief** is necessary because the full-scale **CommunicateBelief** is not an option. We analyze the consequences of such regional communication in Corollary 8.13.

Given an LJF $\mathcal{F}$, if we recursively remove some terminal hypernodes, then those that remain from a subhypertree $\mathcal{F}'$. The hypernodes in $\mathcal{F}'$ are connected, and for each hypernode, its adjacency is identical to that in $\mathcal{F}$. The subhypertree captures the intuitive notion of a region. The following corollary shows that regional communication is exact relative to the observations contained in the region and relative to the knowledge contained in the entire LJF.

Corollary 8.13 *Let $\mathcal{F}$ be a globally consistent LJF representation over V and $P(V)$ be the JPD. Let $\mathcal{F}'$ be a subhypertree of $\mathcal{F}$. After some agents with local observations in $\mathcal{F}'$ have performed **EnterEvidence** locally, a **CommunicateBelief** is performed in $\mathcal{F}'$ only. Then for each cluster C of each local JT in $\mathcal{F}'$,*

$$B_C(C) = \text{const} * \sum_{V \setminus C} P(V|\mathbf{x}),$$

where the configuration $\mathbf{x}$ denotes all the observations entered in $\mathcal{F}'$.

Proof: Apply Theorem 8.12 to the case in which only agents on $\mathcal{F}'$ perform **EnterEvidence**. The corollary follows. □

8.10 Alternative Methods for Multiagent Inference

We have organized probabilistic knowledge of multiple agents as an MSBN, compiled an MSBN into an LJF representation, and performed belief updating using concise message passing in the LJF. This framework for cooperative multiagent inference is extended from the junction-tree-based inference method for single-agent BNs (Chapters 3 through 5). In Sections 2.7 and 2.8, we briefly introduced two alternative inference methods for BNs, loop cutset conditioning and forward sampling (as a typical example of stochastic simulation). Can these methods be extended for multiagent inference in MSBNs?

Consider extending the loop cutset conditioning to MSBNs. Recall that the method converts a multiply connected BN into multiple tree-structured BNs, performs $\lambda - \pi$ message passing in each of them, and then combines the results to obtain posteriors. The key step of the method is to observe a set of nodes hypothetically, the loop cutset, in the multiply connected BN so that all cycles can be broken.

In an MSBN, cycles can exist both within local DAGs and across multiple local DAGs. Assume that a total of k cycles exist in the DAG union. Finding a loop cutset that can break the k cycles is the first challenge, and it requires a distributed search. After such a cutset is found, the multiply connected DAG union needs to be converted to $O(2^k)$ tree-structured DAG unions, and concise message passing needs to be performed in each of them. Because communication within each tree-structured DAG union requires coordination among all agents involved, the communication for the entire multiagent system essentially requires such coordination to be repeated $O(2^k)$ times – once for each tree-structured DAG union. In other words, the amount of communication is $O(2^k)$ times as much as that needed by a single tree-structured DAG union. As discussed in Section 1.4, the agent autonomy dictates that constant communication among agents is undesirable or unavailable in a multiagent system. Even if $O(2^k)$ communications are acceptable, because the loop cutset must assume a different, hypothetically observed configuration at each of the $O(2^k)$ tree-structured DAG unions, the adequate assignment of a configuration to each DAG union and the coherent combination of partial beliefs from each DAG union without a centralized control appear to require complex coordination and communication among agents. In essence, the extended loop cutset conditioning appears to demand much more communication and complex coordination among agents than are required by message passing in LJFs.

Next, consider extending forward sampling to MSBNs. Recall that the method randomly generates the value of each node in a multiply connected BN given the values of its parent nodes according to the conditional probability distribution stored in the node. A large number of configurations are thus generated, and the posteriors are approximated by frequency counting. Because a value for each node is passed to each child during the generation of each configuration, a massive volume of messages needs to be passed over the DAG structure.

According to forward sampling, the value of a child node is decided after the values of all its parents have been determined. In an MSBN, the parents of a node, however, may be located in an adjacent local DAG. Hence, for each simulated configuration, messages must be passed between agents to complete the configuration. One important difference from message passing in LJFs is worth analyzing. In an LJF, all potentials (one for each linkage) for the same linkage tree can be passed from one agent to another in one batch, and two batches (one in each direction of

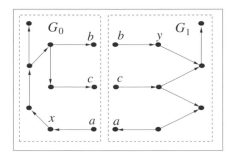

Figure 8.17: Two adjacent local DAGs in an MSBN.

the hyperlink) are sufficient to complete the communication. For extended forward sampling, the situation is different:

An agent A_0 may contain a variable x, but the values of its parent variables $\pi(x)$ must be determined by another agent A_1 because the parents of $\pi(x)$ are contained only in A_1. Thus, A_0 needs to wait until A_1 sends the values of $\pi(x)$. It is also possible that A_1 contains a variable y, the parents of $\pi(y)$ are contained only in A_0, and x is an ancestor of y. Hence, A_1 cannot determine the values for all variables it shares with A_0 and as a result cannot send them in one batch. Instead, A_1 must wait until A_0 sends the values of $\pi(y)$. Figure 8.17 illustrates the situation in which A_0 must wait for the value of a from A_1, and A_1 must wait for the value of b from A_0. Because the DAG union of an MSBN is acyclic, a deadlock is not possible. However, if a directed path crosses the interface between two agents k times, then messages must be passed back and forth k times between the two agents before all variables on the path settle on their values. Note that a directed path may pass across multiple agent interfaces. Hence, during the generation of each configuration, the number of batches of messages between each pair of adjacent agents is upper bounded by the number (which is usually much larger than 2) of d-sepnodes between them. Because each batch of messages incurs a communications cost independent of its volume, it is undesirable to exchange a large number of batches.

If the configurations are generated one by one, as in the single-agent case, the aforementioned communications cost will have to be multiplied by the number of configurations. To reduce this cost, the sequential generation of configurations as normally performed in a single-agent BN should be avoided. Agents can generate a large number of values for each local variable before a batch of messages is exchanged. For example, A_1 may generate 1000 values for a and pass all of them to A_0 and later receive 1000 values for each of b and c. The price to be paid is that all 1000 values for each variable must be saved in some way until the compatibility of each configuration with observations is resolved. Note that if 1000 configurations do not produce approximate posteriors that are sufficiently accurate, either the batch size (1000) can be increased or additional batches can be exchanged.

After configurations have been generated, another problem that must be resolved is to determine which of them are compatible with the observations. This can be achieved by letting each agent label each configuration that is incompatible with its local observation. All such labels must then be communicated among all agents to weed out each configuration that is labeled as incompatible by any agent.

Overall, extended forward sampling appears to be more manageable than extended loop cutset conditioning. In comparison with the message passing in LJFs, the per-batch-volume of interagent messages tends to be larger owing to the number of configurations to be generated, and more batches need to be exchanged because of the need to follow causal ordering of variables and to propagate configuration labels.

In summary, the two alternative methods do not necessitate constructing the LJF representation, although extended loop cutset conditioning requires distributed search for the loop cutset and local DAG conversion. In general, the on-line communication cost of the two methods appears significantly higher than that of message passing in LJFs. Because the LJF representation is constructed off-line, message passing in LJFs during on-line inference has a simpler coordination and incurs a lower communications cost.

8.11 Bibliographical Notes

Efforts to improve the inference efficiency in MSBNs have been ongoing since the initial proposal of the framework. The first set of inference algorithms was presented in Xiang [80] and Xiang et al. [96]. The algorithms were designed under the single-agent paradigm. An algorithm **ShiftAttention** directs the computation to a single subnet at any time according to the subdomain of the user's current focus of attention (see Exercise 6). The algorithm **UpdateBelief** at the time requires multiple performance of **UnifyBelief** – once for each linkage. An improved version of **UpdateBelief** was later proposed by Xiang [83] and reduces the total number of clusters in the local JT to be processed by coordinating each **UnifyBelief** to be performed.

As the MSBN framework was extended to the multiagent paradigm (Xiang [84]), so must the inference algorithms be extended to address asynchronous and parallel observations by multiple agents. The algorithm **CommunicateBelief** was then proposed. Optimizing message passing by taking advantage of the semiparallel pattern of communication (as demonstrated in Section 8.9) was also analyzed. The algorithms presented in this chapter correspond to a more recent improvement (Xiang [88]). By using the extended linkage belief, the improved algorithm **UpdateBelief** reduces the number of times **UnifyBelief** is to be performed to two.

Lazy inference in MSBNs by Xiang and Jensen [91] provides an alternative representation and inference formalism to the LJF representation discussed in this chapter. It uses a factorized representation for each cluster potential and multiple local JTs for each agent. Lazy inference can improve space complexity by factorizing potentials.

To address the knowledge engineering issues in building large BN-based systems, the object-oriented Bayesian networks (OOBNs) have been proposed (Koller and Pfeffer [35]). An OOBN is intended to function under the single-agent paradigm. Belief updating in OOBNs can be performed by using **CommunicateBelief**, or by using single-agent-oriented MSBN inference such as **ShiftAttention** (see Exercise 6), or converting the OOBN to a BN and applying any BN inference method (Bangso and Wuillemin [1]). The last two options are not applicable to a multiagent MSBN system owing to the requirements of agent autonomy and privacy.

Equipment troubleshooting using MSBNs was demonstrated by Xiang and Geng [90]. Excrcise 1 is an extension of a result by Dawid and Lauritzen [12].

8.12 Exercises

1. Prove the following theorem:

   Let $T = (V, \Omega, E)$ be a junction tree, where each cluster Q_i is associated with a probability distribution $f_{Q_i}(Q_i)$ and each separator S_j is associated with a probability distribution $f_{S_j}(S_j)$ such that T is locally consistent.

   Then there exists a unique probability distribution

   $$P_T(V) = \frac{\prod_i f_{Q_i}(Q_i)}{\prod_j f_{S_j}(S_j)}$$

   such that
   (a) for each cluster Q_i, $\sum_{N \setminus Q_i} P_T(V) = f_{Q_i}(Q_i)$, and
   (b) T is an I-map of $P_T(V)$.

2. Let the subdomains $\{V_i\}$ of a total universe V be organized into a hypertree F (Definition 6.8). Let V be populated by a set of cooperative agents $\{A_i\}$, each of which holds its belief over V_i expressed as a probability distribution $P_i(V_i)$ and every two of adjacent agents in F are consistent in their beliefs over their interface.

   Using the theorem in Exercise 1, justify that the joint system potential of a globally consistent LJF representation is a coherent joint belief of all agents.

3. In Section 8.8, it is assumed that the shared variables between a pair of agents are a small subset of either subdomain involved. What is the consequence if this assumption is not true?

4. A given agent's belief about its subdomain is usually more sensitive to the probabilistic information contained in agents close to it on the hypertree than those located far away. Under what condition is this not true?

5. Construct an experimental MSBN and perform multiagent belief updating using an automated toolkit such as WebWeavr-III.

   Convert the MSBN into a centralized BN by obtaining the union of all subnets. Use any exact method and the same set of observations above for belief updating. Compare the posteriors obtained in each case.

6. Suppose that in an MSBN-based multiagent system, multiple agents make observations but exactly one agent observes at any time. The order in which agents observe is unknown in advance. The belief of the observing agent must be updated as soon as the observation is made. An observation may involve multiple variables. Design an algorithm to implement this requirement. (Hint: You may use any algorithms presented in this chapter except **CommunicateBelief**. This algorithm is equivalent to **ShiftAttention** in the single-agent paradigm.)

9

Model Construction and Verification

In Chapter 7, we developed algorithms to compile a given MSBN into an LJF, and in Chapter 8, we described multiagent belief updating by message passing. We have not discussed how an MSBN is constructed. In this chapter, we address issues related to the construction of MSBNs. In particular, we study how to integrate an MSBN-based multiagent system from agents developed by independent vendors. As detailed in Chapter 6, an MSBN is characterized by a set of technical conditions. Because independently developed subnets may not satisfy these conditions collectively, automatic verification is required to avoid the "garbage in and garbage out" scenario. The verification process becomes subtle when agents are built by independent vendors and the vendors' know-how needs to be protected. We develop a set of distributed algorithms that verify the integrity of an integrated multiagent system against the technical requirements of an MSBN.

The motivations for multiagent distributed verification are presented in Section 9.2. Section 9.3 addresses the verification of subdomain division among agents. The concept of the interface graph is introduced to capture the knowledge of the system integrator about the public aspects of multiple agents. A simple method to verify the subdomain division using the interface graph is presented. Constructing the hypertree of the MSBN at the logical level once the subdomain division is validated is also described. After the logical hypertree is constructed by the system integrator, each agent needs to be notified of its location in the hypertree and the physical addresses of its adjacent agents, which will be used for subsequent interaction. This process is termed *registration* and is presented in Section 9.4. After registration, the hypertree is physically constructed. All subsequent verification, compilation, and inference operations can be performed along the hypertree. Although each subnet has a local DAG dependence structure, the union of all such local DAGs may still be cyclic. Section 9.5 develops a set of distributed algorithms that allow multiple agents to cooperate and verify the acyclicity of the DAG union. Section 9.6

presents a set of distributed algorithms that verify the agent interface, the d-sepset condition.

9.1 Guide to Chapter 9

This chapter considers how to construct a multiagent system and to verify that it satisfies the technical conditions of the MSBN representation. Section 9.2 discusses the motivations for integrating an MSBN system from agents developed by independent vendors. Given such a set of agents, it is not necessarily the case that they collectively meet the requirements of an MSBN, as elaborated in Chapter 6. To protect agent privacy, integration and verification need to be performed without revealing the internal details of each agent.

Section 9.3 deals with the first issue in construction and verification: the partitioning of the total universe into subdomains by a set of given agents. For instance, suppose that four (trivial) agents have the following subdomains:

$$V_0 = \{a, b, w, x\}, \quad V_1 = \{c, d, x, y\}, \quad V_2 = \{e, f, y, z\}, \quad \text{and} \quad V_3 = \{g, h, w, z\}.$$

Note that the variables w, x, y, and z are shared among the agents whereas variables a, b, c, d, e, f, g, and h are private. Hence, what is actually known to the system integrator is only the following:

$$W_0 = \{w, x\}, \quad W_1 = \{x, y\}, \quad W_2 = \{y, z\}, \quad \text{and} \quad W_3 = \{w, z\}.$$

One of the requirements of an MSBN is that the subdomains be organized into a hypertree. Hence, the issue is to decide whether a hypertree exists for V_0 through V_3 given only the knowledge of W_0 through W_3. Section 9.3 devises a graphical model called the *interface graph* that encodes the knowledge on W_0 through W_3. Using the interface graph and some graphical manipulation, the existence of a hypertree can be determined correctly without knowing the private variables. If a hypertree does exist, the method can also produce the topology of the hypertree.

Once the topology of the hypertree is determined, each agent needs to be informed how to contact its neighboring agents on the hypertree through computer network connections. That is, each agent needs to know the Internet address or the universal resource locator (URL) or something similar for each adjacent agent so that it can pass messages to the agent. The process is termed *registration* and is elaborated in Section 9.4. After registration, the multiagent system is up and running through the hypertree backbone.

Another important requirement of an MSBN is that the union of local DAGs of agents should be a DAG. It turns out that a set of local DAGs does not necessarily union into a DAG. An example is shown in Figure 9.1 in which the hypertree is a hyperstar with G_3 at the center. The interface between G_0 and G_3 is $\{n, k\}$. The

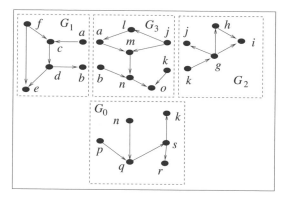

Figure 9.1: Four local DAGs.

interface between G_1 and G_3 is $\{a, b\}$, and that between G_2 and G_3 is $\{j, k\}$. Each local graph is a DAG. However, a directed cycle exists across all four local DAGs. The cycle cannot be detected by examining each pair of adjacent local DAGs either.

Centralizing all local DAGs to a single agent for cycle detection violates agent privacy. Section 9.5 develops a cycle detection method based on message passing. Each message concerns only a node shared by agents and whether such a node has any parent or child in each agent. The message neither reveals the number of parents or children in each agent nor what they are. Using such messages and some control messages, agents are able to cooperate and determine whether the DAG union is acyclic.

Still another important requirement of an MSBN is that each agent interface should be a d-sepset. As defined in Chapter 6, an agent interface is a d-sepset if each shared node in the interface is a d-sepnode. A shared node is a d-sepnode if a local DAG contains its parents in all local DAGs. Therefore, whether a shared node is a d-sepnode cannot be determined by examining any one interface and related local DAGs. As a consequence, whether an agent interface is a d-sepset cannot be determined only by the two agents involved either. To verify the d-sepset condition, agents must cooperate.

Section 9.6 develops a suite of algorithms for d-sepset verification. The algorithms are also based on message passing. A message concerns only a node shared by agents and whether an agent contains parents not shared by other agents (private parents). The message does not reveal how many private parents the agent has nor what they are. Through communicating these and other messages, agents can determine correctly whether all agent interfaces in a hypertree DAG union are d-sepsets.

9.2 Multiagent MSBN System Integration

We consider how to construct a cooperative multiagent system based on the MSBN representation. It is possible that all agents are developed by a single vendor (which could be a single person or a firm). More commonly, agents are developed by independent vendors and are integrated into a system by an integrator. Consider a system for monitoring and troubleshooting a piece of complex equipment. A modern piece of equipment is commonly assembled from components manufactured by different suppliers. The supplier of each component has knowledge about the internal functional structure of the component and is in a good position to construct the agent for monitoring the component. The vendor of that equipment assembles the components. Because the vendor has knowledge of the structure by which the components are related to each other and about the interface between components, it is also in a good position to assemble the component agents into an MSBN-based system.

The primary task of integration, given a set of agents and their interfaces, is to determine the hypertree topology and the location of each agent on the hypertree. We refer to this process as the *logical* construction of the hypertree. Because the agents are likely executing at remote locations over a geographical area and connected through a local area network (LAN) or a wide area network (WAN), each agent needs to know which agents on the hypertree are adjacent and how to communicate with them. That is, after the hypertree is constructed logically, each agent needs to be notified of its logical neighbors on the hypertree and their physical addresses in the LAN or WAN. We refer to this process as the *physical* construction of the hypertree. Once the hypertree is physically constructed, a multiagent system is established, and direct and indirect interaction among agents can take place. We deal with the logical construction of the hypertree in the next section and address the physical construction in Section 9.4.

For the multiagent system to function correctly with respect to the performance properties we established in Chapter 8, the system must satisfy the conditions specified in Definition 6.15. Independently developed agents do not always integrate into a valid MSBN. Certain conditions, such as the hypertree requirement (Definition 6.8), need to be verified during the logical construction of the hypertree. Afterwards, a physically established multiagent system may still violate additional technical conditions of an MSBN. Hence, verification of these conditions is necessary before any compilation and inference can be performed. In Sections 9.5 and 9.6, we consider automatic verification of the acyclicity of the DAG union and the d-sepset agent interface.

As an operational constraint, we are primarily interested in distributed algorithms that admit agent privacy. That is, the construction and verification process should not reveal the internal structure of each agent. The reasons are similar to those we have argued in Chapter 7 in the context of distributed compilation: Distributed and privacy-protecting verification protects agent vendors' know-how and facilitates dynamic formation of a multiagent MSBN. It would be inconsistent to insist on agent privacy for compilation and inference but to disregard privacy in verification.

9.3 Verification of Subdomain Division

A set of agents and their associated subnets define a division $\{V_0, \ldots, V_{n-1}\}$ of the total universe $V = \cup_i V_i$ into subdomains. According to Definition 6.15, subnets should be organized as a hypertree. Definition 6.8 defines the hypertree in terms of a set of subgraphs. If we ignore the graphical structure of each subgraph and focus on the relation between subsets V_i, then Definition 6.8 dictates a connected tree Ψ in which each node is labeled by a V_i and each link between V_k and V_m is labeled by $V_k \cap V_m$ such that for each i and j, $V_i \cap V_j$ is contained in each node on the path between V_i and V_j. Recall the definition of junction tree in Section 3.7. Clearly, if we treat the set $\{V_0, \ldots, V_{n-1}\}$ of subdomains as a set of clusters, then Ψ is in fact a junction tree.

In Section 4.5, it has been shown that, given an arbitrary undirected graph, a junction tree whose clusters are the cliques of the graph may not exist. Hence, given an arbitrary division of the total universe into a set of subdomains, a corresponding hypertree may not exist.

More specifically, recall from Theorem 4.10 that, given an undirected graph G, a junction tree can be constructed using the cliques of G as clusters if and only if G is chordal. Given the set $\{V_0, \ldots, V_{n-1}\}$, a graph G can be constructed with the set $V = \cup_i V_i$ as the nodes. Using the algorithm **IsChordal** in Section 4.6, whether G is chordal can be determined, and hence whether a hypertree exists out of the subdomain division $\{V_0, \ldots, V_{n-1}\}$ can be determined. However, such a verification test requires the knowledge of each V_i. For each agent A_i, V_i can be partitioned into two subsets of variables. Denote the set of all variables to be shared with other agents by $W_i \subset V_i$. We call W_i the *public* variables, and $V_i \setminus W_i$ the *private* variables of agent A_i. Because the above test requires disclosure of the private variables, the commitment to agent privacy prevents such a test.

In the following algorithm, we present a method that does not violate agent privacy. We designate the verification task to the system integrator. We assume that the integrator has the knowledge of all public variables but not for private variables. Using public knowledge, the integrator can build a graph, which we refer to as the *interface graph*.

Algorithm 9.1 (GetInterfaceGraph) *Given the collection of sets of public variables $\{W_0, \ldots, W_{n-1}\}$, create a new set $X = \{x_0, \ldots, x_{n-1}\}$ such that $X \cap (\cup_i W_i) = \emptyset$. Define a new collection $\{W'_0, \ldots, W'_{n-1}\}$ where $W'_i = W_i \cup \{x_i\}$. Create an undirected graph G whose nodes $W' = \cup_i W'_i$. The links in G are such that each W'_i is complete and no other links exist. Return G.*

Figures 9.2 and 9.3 illustrate **GetInterfaceGraph** using a trivial total universe. In Figures 9.2(a), the total universe is divided into five subdomains. The public

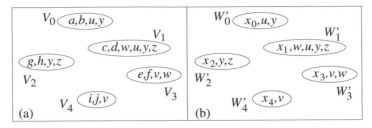

Figure 9.2: Subdomain division verification.

variables are

$$W_0 = \{u, y\}, \quad W_1 = \{w, u, y, z\}, \quad W_2 = \{y, z\}, \quad W_3 = \{v, w\}, \quad \text{and} \quad W_4 = \{v\}.$$

Note that the integrator does not have the knowledge of (a). Instead, only the preceding list is accessible. The integrator defines $\{W_0', \ldots, W_4'\}$, as shown in (b). The interface graph thus created is shown in Figures 9.3.

The following theorem shows that whether or not a junction tree can be constructed from the subdomains $V_0, \ldots, V_{n-1}$ can be determined according to whether the interface graph is chordal.

Theorem 9.1 *Let* $\{V_0, \ldots, V_{n-1}\}$ *be a collection of sets and* $\overline{V_i} = \cup_{\substack{j=0 \\ j \neq i}}^{n-1} V_j$ *(i =* $0, \ldots, n-1$*) such that* $W_i = V_i \cap \overline{V_i} \neq \emptyset$ *and* $V_i \backslash \overline{V_i} \neq \emptyset$ *(i = 0, \ldots, n-1). Let* G *be the interface graph created from the collection* $\{W_0, \ldots, W_{n-1}\}$. *Then a junction tree exists with* $V_0, \ldots, V_{n-1}$ *as clusters if and only if* G *is chordal.*

The first restriction $V_i \cap \overline{V_i} \neq \emptyset$ says that each V_i has a nonempty intersection with at least another set. The second restriction $V_i \backslash \overline{V_i} \neq \emptyset$ says that each V_i has some elements that are unique. These restrictions reflect the properties of a division of a total universe into subdomains in a multiagent system. We prove the theorem below.

Proof: Suppose G is chordal. Each W_i' is a distinct clique in G because it contains a unique node x_i. There are exactly n cliques in G, $W_0', \ldots, W_{n-1}'$, due to the

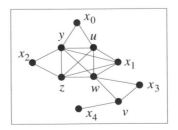

Figure 9.3: Testing the existence of a hypertree.

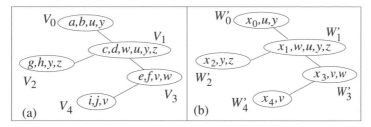

Figure 9.4: Determining hypertree topology from public variables only: (a) A junction tree constructed from the subdomain cluster of Figure 9.2; (b) An isomorphic junction tree constructed using public variables only.

construction of G. Given that G is chordal, a junction tree T' exists whose clusters are cliques of G by Theorem 4.10. That is, each cluster of T' is labeled by a W_i'. We construct a tree T that is isomorphic to T'. Each cluster of T is, however, labeled by V_i. For any two clusters in T, the intersection $V_i \cap V_j = W_i \cap W_j$. Hence, T is also a junction tree.

Next suppose a junction tree T exists with $V_0, \ldots, V_{n-1}$ as clusters. From T, we construct a tree T' isomorphic to T with each cluster labeled by W_i'. Using the argument above, T' is a junction tree, and hence a corresponding undirected graph G is chordal by Theorem 4.10. □

Applying Theorem 9.1 to the example in Figure 9.2, we can test whether the interface graph in Figure 9.3 is chordal using the algorithm **IsChordal** (Section 4.6). The test is positive. Therefore, there exists a junction tree constructed from the subdomain division, as shown in Figure 9.4(a). In fact, the topology of the junction tree can be determined by a junction tree created using $W_0', \ldots, W_{n-1}'$ as clusters, as shown in Figure 9.4(b). Because the junction tree in Figure 9.4(b) and that in (a) are isomorphic, the integrator can use the topology obtained from $W_0', \ldots, W_{n-1}'$ to organize the agents $A_0, \ldots, A_{n-1}$ into a hypertree. We say that the hypertree is now logically constructed.

As another example, consider the subdomain division in Figure 9.5(a). Here the variable h in V_2 becomes a public variable shared by V_4. The new collection $\{W_0', \ldots, W_4'\}$ is shown in (b), and the corresponding interface graph is shown in (c). The graph is nonchordal because $\langle h, v, w, z, h \rangle$ forms a chordless cycle. Hence, the subdomain division cannot yield a hypertree agent organization.

9.4 Agent Registration

Once the integrator has verified the subdomain division and determined the hypertree topology, the verification proceeds to the other requirements of an MSBN.

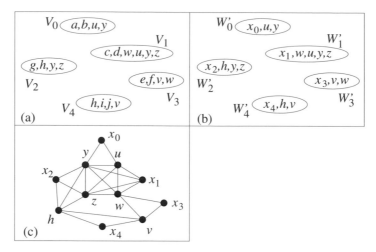

Figure 9.5: A subdomain verification with a negative result: (a) Each cluster variables in each subdomain are substituted; (b) Private variables in each subdomain are substituted by a variable x_i; (c) The corresponding interphase graph.

As we have seen, the hypertree structure provides a backbone for effective exchange of information among agents during compilation (Chapter 7) and inference (Chapter 8). We will see that it is also sufficient to support distributed verification. To this end, each agent needs to be notified of its neighbors on the hypertree for future exchange of information. In other words, before agents can participate in the activities in the system, it is necessary for each agent to locate those agents physically in order to communicate in the future. For example, if agents are remotely located in the Internet, each agent needs to know the Internet address of each other adjacent agent on the hypertree. In addition, the agent needs to know the variables shared with each adjacent agent. We refer to this process as *registration*, which is commonly used to refer to the similar processes in distributed systems (Tanenbaum [73]).

As discussed in Section 9.3, before registration the system integrator has determined the hypertree topology as well as the set of variables (agent interface) shared between each pair of adjacent agents. The registration can be implemented by a *registrar* agent on behalf of the system integrator. The registrar has access to the following information:

1. The set of agents in the system. We assume that each agent is known to the registrar by a unique logical name (e.g., "boiler 3 guardian" in a chemical plant monitoring system).
2. The topology of the hypertree and the location of each agent on the hypertree.
3. The agent interface between each pair of adjacent agents.

Each agent is given the physical address of the registrar. The registration can then proceed as follows:

Algorithm 9.2 (Registration)

1. Each agent sends the registrar its logical name and physical address.
2. The registrar collects the addresses until all are received.
3. For each agent A_i, the registrar sends A_i the physical address of each adjacent agent and the agent interface.

We have assumed that each agent is known to the integrator (and registrar) by a logical name instead of its physical address. The actual physical address is not known to the relevant agents until registration. This arrangement allows flexible physical location and relocation of agents, which is known as *location transparency* in distributed systems (Tanenbaum [73]).

After registration, each agent knows its hypertree neighbors and the corresponding agent interface. The hypertree is said to be physically constructed.

9.5 Distributed Verification of Acyclicity

9.5.1 The Issue of Acyclicity Verification

According to Definition 6.15, the structure of an MSBN, denoted by $G = \sqcup_i G_i$, where each G_i is a local DAG, is a hypertree MSDAG. As specified in Definition 6.13, in a hypertree MSDAG, not only is each local subgraph a DAG, but the union G is also a DAG (acyclic). We refer to $G = \sqcup_i G_i$ as a hypertree DAG union because the task here is to determine whether G is acyclic.

Figure 9.6 shows two local DAGs G_1 and G_2 with the agent interface $\{a, b\}$. However, the DAG union contains a directed cycle $\langle a, c, d, b, g, a \rangle$ and thus is *cyclic*. The cycle can be detected if acyclicity is tested on the DAG union. Although such pairwise verification may detect some directed cycles, the pairwise acyclicity in a hypertree DAG union does not guarantee the global acyclicity.

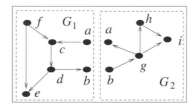

Figure 9.6: A cyclic DAG union.

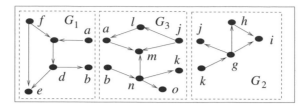

Figure 9.7: The three local DAGs are pairwise acyclic. However, their union is cyclic.

Consider the three local DAGs in Figure 9.7. The union of G_1 and G_3 is acyclic and so is the union of G_3 and G_2. However, when the union of the three DAGs are obtained, a directed cycle $\langle a, c, d, b, n, k, g, j, l, a \rangle$ is formed. This example can be extended so that the union of $k \geq 2$ local DAGs is acyclic but the union of $k + 1$ local DAGs is cyclic. Hence, a distributed verification of acyclicity requires cooperation beyond pairs of agents.

9.5.2 Verification by Marking Nodes

We show below that acyclicity of a directed graph can be verified by marking root and leaf nodes recursively. Once this is established, agents can mark private nodes (variables) locally and mark shared nodes by cooperation.

A node x is *marked* if x and arcs connected to x are ignored during the further verification process. Lemma 9.2 says that marking a root or a leaf does not change acyclicity.

Lemma 9.2 *Let G be a directed graph and x be either a root or a leaf in G. Then the acyclicity of G remains unchanged after x is marked.*

Proof: If G is acyclic, then marking x cannot create a directed cycle in G. Suppose G is cyclic. Then there exists a nonempty set O of directed cycles in G. If x is a root, it does not have any incoming arc. If x is a leaf, it does not have any outgoing arc. Therefore, x cannot participate in any cycles in O, which implies that none of the cycles in O will be changed after x is marked. □

Once a root or leaf is marked, other nodes may become roots or leaves. Hence, marking roots and leaves can be performed recursively while preserving the acyclicity. The following proposition says that if a directed graph is acyclic, every node in it will be marked by recursive applications of Lemma 9.2. On the other hand, if it is cyclic, at least three nodes will be left unmarked.

Proposition 9.3 *Let G be a directed graph. The graph is acyclic if and only if it is empty after recursive marking of roots and leaves.*

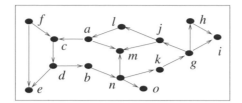

Figure 9.8: A directed graph.

Proof: Without losing generality, we assume that G has at least three nodes and is connected. Suppose G is acyclic. Then G has at least one root and one leaf. According to Lemma 9.2, after all of them are marked, the resultant graph is still acyclic and has at least two less unmarked nodes. Because G has a finite number of nodes, after recursive marking of roots and leaves, eventually G will have no unmarked nodes left.

Next, suppose G is cyclic. Then G has at least one directed cycle θ consisting of at least three nodes. For each node x in θ, x is neither a root nor a leaf and thus cannot be marked as such. Marking of any nodes outside θ cannot turn x into a root or a leaf. Hence, none of the nodes in θ can be marked by recursive marking of roots and leaves. Because G has a finite number of nodes, after recursive marking of roots and leaves eventually there will be no roots or leaves to mark in G whereas all nodes (at least three) in θ are unmarked. □

As an example, consider the directed graph in Figure 9.8. After marking the root node f and the leaf nodes e, i, m, o, the graph in Figure 9.9(a) is obtained. The node h becomes a new leaf. After marking h, the graph in (b) is obtained. There are nine unmarked nodes in (b), but none of them is a root or a leaf. Hence, the original graph in Figure 9.8 is cyclic.

As another example, suppose the arc (g, j) in Figure 9.8 is replaced by the arc (j, g), which results in the graph in Figure 9.10. After marking the root nodes f, j, and the leaf nodes e, i, m, o, the graph in Figure 9.11(a) is obtained. After marking the new root l and the new leaf h, the graph in (b) is obtained. Note that the unmarked nodes form a directed path (a, c, d, b, n, k, g) with a new root a and

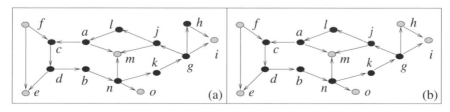

Figure 9.9: Testing acyclicity by node marking. Marked nodes are shown in grey.

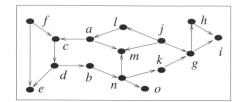

Figure 9.10: Another directed graph.

a new leaf g. By recursively marking the root node and the leaf node, eventually the only unmarked node left will be b. As shown in (c), b is both a new root (no unmarked parents) and a new leaf (no unmarked children), it can be marked to yield the graph in (d). Because no unmarked node is present, the original graph in Figure 9.10 is acyclic.

9.5.3 Marking Nodes in a Hypertree DAG Union

We now consider a hypertree DAG union G in which nodes in each local DAG are partitioned into private nodes and public nodes. Because G is a connected directed graph, one may attempt to apply Proposition 9.3 in order to test its acyclicity. However, although private roots and leaves can be recognized locally within each local DAG, public roots and leaves can only be recognized through cooperation among agents. Consider the DAG union in Figure 9.7 reproduced as Figure 9.12. The node i (private) is a leaf both in G_2 (appearing markable locally) and in the DAG union (hence markable globally). On the other hand, k (public) is a leaf in G_3 (appearing markable locally), a root in G_2 (appearing markable locally), but a

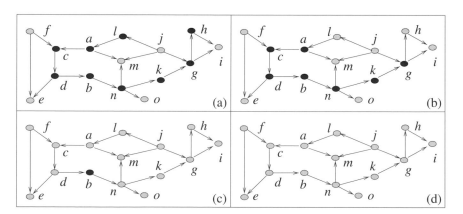

Figure 9.11: Testing acyclicity of a DAG: (a) After marking root and leaf nodes; (b) After marking the new root i and the new leaf h; (c) After recursively marking root and leaf nodes, the only unmarked node left will be b; (d) All nodes are marked.

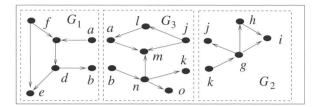

Figure 9.12: Three DAGs which are pairwise acyclic but whose union is cyclic.

nonroot and nonleaf in the DAG union (hence not markable globally). Moreover, marking of public roots and leaves may turn some private nodes into new roots or leaves. For instance, if the arc (g, j) is replaced by (j, g), then j would be a public root. Marking of j would turn l to a new root and make it markable.

The following proposition shows that recursive and alternate marking of private roots and leaves and public roots and leaves is sufficient to determine the acyclicity of a hypertree DAG union.

Corollary 9.4 *Let G be a DAG union. Let G′ be the graph resulting from recursive and alternate marking of private roots and leaves and public roots and leaves in G until no more nodes can be marked. Then G is acyclic if and only if G′ is empty.*

Proof: According to Proposition 9.3, if G is acyclic, at each round of recursive marking, either some private roots and leaves or some public roots and leaves can be marked until G is empty. If G is cyclic, at each round of recursive marking, either some private roots and leaves or some public roots and leaves can be marked until only nodes in directed cycles in G are left unmarked (at least three).

The alternate marking of public and private nodes is necessary. Otherwise, the marking may halt prematurely even if G is acyclic. □

To illustrate the necessity of alternate marking, consider the acyclic DAG union in Figure 9.13. Without using alternate marking, two options exist:

1. Recursive marking of all private roots and leaves followed by recursive marking of all public roots and leaves, or

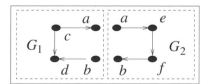

Figure 9.13: An acyclic DAG union.

2. Recursive marking of all public roots and leaves followed by recursive marking of all private roots and leaves.

With the first option, private nodes c (root) and d (leaf) in G_1 will be marked in the first stage. Neither of the private nodes e and f in G_2 can be marked at this stage. In the second stage, the public nodes a (now a root) and b (now a leaf) can be marked. The marking terminates with e and f unmarked.

With the second option, no public nodes can be marked in the first stage because neither a nor b is a root or leaf. In the second stage, private nodes c (root) and d (leaf) in G_1 will be marked. The marking terminates with a, b, e, and f unmarked.

If alternate marking is performed by starting with private nodes, c and d will be marked in the first stage. In the second stage, public nodes a and b will be marked. In the third stage, private nodes e and f will be marked. Now the resultant graph is empty. Alternate marking by starting with public nodes gives the same result.

Note that in order to recognize a public root shared by two agents, one agent must tell the other that there are no unmarked parents in its local DAG. The same is true for recognition of a public leaf. Hence, marking public nodes requires cooperation among agents.

Note also that Corollary 9.4 does not require G to satisfy the hypertree condition even though our application is under the context of a hypertree DAG union. Corollary 9.4 forms the basis for a distributed verification algorithm, which we present next.

9.5.4 Multiagent Verification

As demonstrated in Sections 9.5.1 and 9.5.3, in order to verify the acyclicity of a hypertree DAG union, agents must cooperate. We present a set of algorithms, most of which are recursive, to be executed by an agent. We follow the same convention used in earlier chapters for naming agents: An algorithm executed by an agent is activated by a call from an entity known as the caller. We use A_0 to denote the agent called to execute the algorithm. The caller is either an adjacent agent of A_0 in the hypertree or the system coordinator, which could be either a human or an agent. If the caller is an adjacent agent, we denote it by A_c. If A_0 has adjacent agents other than A_c, we denote them by $A_1, \ldots, A_k$. The subdomains of $A_c, A_0, \ldots, A_k$ are $V_c, V_0, \ldots, V_k$, and their subgraphs are $G_c, G_0, \ldots, G_k$, respectively. We denote $V_c \cap V_0$ by I_c and $V_0 \cap V_i$ by I_i ($i = 1, \ldots, k$). If an algorithm is executed by the system coordinator, we will use A_* to describe any agent selected by the system coordinator.

Cooperation requires interaction, which is generally more expensive and less efficient than local computation. Hence, it is desirable to simplify the task for

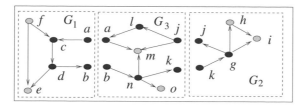

Figure 9.14: The DAG union in Figure 9.7 after local preprocessing.

cooperation as much as possible. According to Corollary 9.4, private roots and leaves in local DAGs can be marked separately and recursively. We define **Pre-Process** below to mark these nodes before cooperation starts.

Algorithm 9.3 (PreProcess) *When **PreProcess** is called on A_0, it does the following:*

1. Agent A_0 recursively marks each private root or leaf in its local DAG G_0.
*2. For each agent A_i ($i = 1, \ldots, k$), A_0 calls A_i to run **PreProcess**.*

PreProcess can be activated at any agent by the system coordinator. The processing will then propagate outwards along the hypertree. The end result of **PreProcess** is independent of the agent that activated the processing. Note that no internal structural information within each agent is revealed to other agents.

As an example, consider the hypertree DAG union in Figure 9.7. Suppose that three agents are organized into the hyperchain $A_1 - A_3 - A_2$. If **PreProcess** is activated at A_1, then it is performed first at A_1, then at A_3, and finally at A_2. Alternatively, **PreProcess** can be activated at A_3 and then be performed at A_1 and A_2. Either way, the DAG union after **PreProcess** is at the same state and is shown in Figure 9.14.

In general, after **PreProcess** is completed by all agents, the nodes that are left unmarked in each local DAG are either nodes that form directed paths terminated at public nodes or isolated public nodes. The first case occurs in this example (e.g., the directed path (a, c, d, b) in G_1 and (b, n, k) in G_3). To see the second case, consider another example in Figure 9.15(a). The only difference of the DAG union from the previous example is the reversal of arc (g, j) in G_2. The end result of **PreProcess** is shown in (b). Two isolated public nodes j and k (each qualifies as a root and as a leaf) are left unmarked in G_2. **PreProcess** prevents marking such public roots and leaves because an isolated public node in a local DAG may still participate in a directed cycle in other local DAGs. It would be inconsistent to mark the node in one local DAG and to keep it unmarked in another. To mark the remaining nodes, further cooperation among agents is needed.

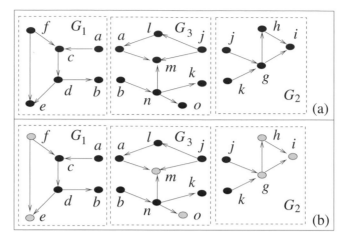

Figure 9.15: Three local DAGs whose union is shown in Figure 9.10: (a) The local DAGs before **PreProcess** is performed. (b) After **PreProcess** is performed.

To find out if a public node x can be marked, an agent cooperates with others and performs **CollectFamilyInfo** to determine if x is a root or leaf in the DAG union. Messages in the form of a triplet (x, p, c) are passed among all agents who share x. The purpose is to collect the parent and child information for x, where p signifies whether any agents have parents of x in their local DAGs and c signifies whether any agents have children of x.

Algorithm 9.4 (CollectFamilyInfo) *When **CollectFamilyInfo(x)** is called on an agent A_0, it does the following:*

1. *Agent A_0 forms a triplet $t_0 = (x, p_0, c_0)$, where $p_0 = 1$ if G_0 contains an (unmarked) parent of x and $p_0 = 0$ otherwise, and $c_0 = 1$ if G_0 contains a (unmarked) child of x and $c_0 = 0$ otherwise.*
2. *If A_0 does not share x with any agent A_i ($i = 1, \ldots, k$), or $p_0 = c_0 = 1$, then A_0 returns t_0 to **caller**.*
3. *Otherwise, A_0 calls **CollectFamilyInfo(x)** in each agent A_i ($i = 1, \ldots, k$) that shares x.*
4. *After each A_i ($i = 1, \ldots, k$) being called has returned their triplets (assuming k agents are called), $t_1, t_2, \ldots, t_k$, A_0 returns a triplet $t = (x, p = \max_{i=0}^{k} p_i, c = \max_{i=0}^{k} c_i)$ to **caller**.*

When **CollectFamilyInfo(x)** is activated in an agent A_0, it will be performed recursively outwards from A_0 to each other agent (except the caller agent) on the hypertree as long as the agent shares x. The outward propagation terminates at each agent A_i that has no other agents to call, either because A_i has no adjacent agent except the caller or because A_i does not share x with any other adjacent agent

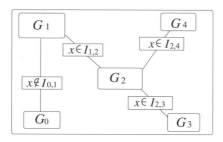

Figure 9.16: Illustration of **CollectFamilyInfo.**

except the caller. The triplet (x, p, c) will be passed from each such A_i inwards towards A_0. For example, suppose that **CollectFamilyInfo(k)** is activated at A_2 in Figure 9.15(b). Agent A_2 forms a triplet $(k, 0, 0)$ and calls **CollectFamilyInfo(k)** in A_3. In response, A_3 forms $(k, 1, 0)$. Because A_3 does not share k with A_1, the outward propagation of **CollectFamilyInfo(k)** stops at A_3. It starts the inward passing of the triplet by returning $(k, 1, 0)$ to A_2. Finally A_2 combines $(k, 1, 0)$ with its own $(k, 0, 0)$ and returns $(k, 1, 0)$ as the final triplet, reflecting the fact that k is a public leaf.

In general, a public node may be shared by several agents that are more widely scattered on the hypertree than in the preceding example. For instance, in Figure 9.16, x is shared by agents A_1 through A_4. Suppose that **CollectFamily-Info(x)** is activated at A_3. If A_3 starts with the triplet $(x, 1, 1)$, then x is neither a public root nor a public leaf. In that case, there is no need to continue collecting family information about x and A_3 will terminate **CollectFamilyInfo(x)** by returning $(x, 1, 1)$ (step 2 of **CollectFamilyInfo**). Otherwise, A_3 will call A_2 to perform **CollectFamilyInfo(x)**. Unless A_2 forms its triplet as $(x, 1, 1)$, it will call A_1 and A_4 to perform **CollectFamilyInfo(x)**. The propagation of **CollectFamilyInfo(x)** stops at A_1 and A_4, and triplets are passed from them to A_2 and then to A_3.

Note that during **CollectFamilyInfo**, the only internal structural information revealed about each agent is whether it contains any parents or children of a public node. The number of parents or children and what they are remain private.

CollectFamilyInfo is efficient, as we will show below. However, it can be improved to perform more efficiently. For instance, when A_0 calls A_i to perform **CollectFamilyInfo(x)**, A_0 may forward its own triplet to A_i. If A_0 has $(x, 1, 0)$ and A_i has $(x, 0, 1)$, then A_i can immediately return $(x, 1, 1)$ to A_0 without calling other agents to **CollectFamilyInfo.** We consider this opportunity in an exercise (see Exercise 2).

Once a public node x is determined to be a root $(p = 0)$ or a leaf $(c = 0)$, the following **DistributeMark** is used to mark the node in every agent that shares it.

Algorithm 9.5 (DistributeMark) *When DistributeMark(x) is called on an agent A_0, it does the following:*

1. *Agent A_0 marks the node x.*
2. *Agent A_0 recursively marks any private root or leaf in its local DAG G_0.*
3. *For each agent A_i $(i = 1, \ldots, k)$ that shares x, A_0 calls DistributeMark(x) in A_i.*

When **DistributeMark(x)** is activated at an agent A_0, it will be performed recursively outwards from A_0 to each agent on the hypertree as long as the agent contains x. Note that **DistributeMark** effectively alternates the marking of a public node with the marking of private nodes, as suggested by Corollary 9.4.

The following **MarkNode** combines **CollectFamilyInfo** and **DistributeMark**. It is used to mark each currently recognizable public root or leaf along the hypertree and to alternate the marking with the marking of private roots and leaves in each agent.

Algorithm 9.6 (MarkNode) *When MarkNode is called on an agent A_0, it does the following:*

1. *Agent A_0 returns **false** if it has no adjacent agent except A_c. Otherwise it continues.*
2. *For each unmarked public node x in G_0, if there exists an adjacent agent A_i $(i = 1, \ldots, k)$ that shares x, A_0 calls **CollectFamilyInfo(x)** on itself. After A_0 returns the triplet (x, p, c), it calls **DistributeMark(x)** on itself if $p = 0$ or $c = 0$.*
3. *Agent A_0 calls **MarkNode** in each A_i $(i = 1, \ldots, k)$.*
4. *If any A_i returns **true** or **DistributeMark(x)** has been called on A_0, then A_0 returns **true** to **caller**. Otherwise, A_0 returns **false** (no node is marked).*

In the second step of **MarkNode**, each public root or leaf contained in A_0 is identified and marked outwards along the hypertree and then any resultant private roots and leaves are also marked. In the third step, the processing continues at adjacent agents of A_0 and outwards on the hypertree. Either in the first step or in the last step, a flag is returned to signify whether any node has been marked at all.

The following **MarkedAll** is used to check whether all nodes in a hypertree DAG union have been marked.

Algorithm 9.7 (MarkedAll) *When MarkedAll is called on an agent A_0, it does the following:*

1. *Agent A_0 returns **false** if there exists an unmarked node in G_0. Otherwise, it continues.*
2. *If A_0 has no adjacent agents except A_c, it returns **true**. Otherwise, A_0 calls **MarkedAll** in each A_i $(i = 1, \ldots, k)$.*

3. *If any A_i returns **false** (with unmarked nodes), then A_0 returns **false**. Otherwise, A_0 returns **true**.*

MarkedAll can be activated at any agent A_0 on the hypertree. It will then be propagated outwards along the hypertree. As soon as an unmarked node is found by any agent, it terminates the propagation of **MarkedAll** and passes a flag of value **false** inwards towards A_0. The flag is then returned by A_0. If no unmarked node is found, **MarkedAll** will be performed by every agent and flags of value **true** will be passed from terminal agents on the hypertree inwards towards A_0. The flag **true** is eventually returned by A_0. According to Corollary 9.4, the DAG union G is acyclic if and only if **true** is returned.

The following top-level algorithm, **TestAcyclicity**, combines the preceding algorithms for cooperative verification of the acyclicity of a DAG union.

Algorithm 9.8 (TestAcyclicity) *Let a hypertree DAG union G be populated by multiple agents with one at each hypernode. The system coordinator does the following:*

1. *Choose an agent A_* arbitrarily.*
2. *Call **PreProcess** in A_*.*
3. *Call **MarkNode** in A_* repeatedly until **false** is returned (no node is marked in the last call).*
4. *Call **MarkedAll** in A_*. If **true** is returned, then conclude that G is **acyclic**. Otherwise, G is **cyclic**.*

9.5.5 Illustration of Cooperative Verification

We illustrate the performance of **TestAcyclicity** with two examples, one with a cyclic DAG union and another with an acyclic DAG union. The first example uses the DAG union in Figure 9.7.

- Suppose A_1 is chosen as A_*. After **PreProcess**, the union is shown as Figure 9.14 and is reproduced as Figure 9.17 for convenience.

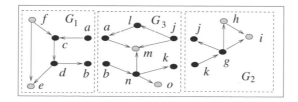

Figure 9.17: Illustration of **TestAcyclicity** in a cyclic DAG union.

- When **MarkNode** is activated in A_1, it calls on itself **CollectFamilyInfo(a)**, which is then propagated to Agent A_3, and returns a triplet $(a, 1, 0)$ to A_1. Subsequently, A_1 generates the final triplet $(a, 1, 1)$ and terminates **CollectFamilyInfo(a)**. Because neither p nor c is zero, **DistributeMark(a)** is not called.

    Agent A_1 then calls on itself **CollectFamilyInfo(b)**, which eventually terminates similarly as **CollectFamilyInfo(a)**.

- Agent A_1 calls **MarkNode** in A_3, which calls on itself **CollectFamilyInfo(j)** and is then propagated to A_2. Agent A_2 returns a triplet $(j, 1, 0)$ to A_3. Subsequently, A_3 generates the final triplet $(j, 1, 1)$ and terminates **CollectFamilyInfo(j)**. No **DistributeMark** is called.

    Agent A_3 then calls on itself **CollectFamilyInfo(k)**, which eventually terminates similarly as **CollectFamilyInfo(j)**.

- Agent A_3 calls **MarkNode** in A_2, which returns **false** because A_2 has no other adjacent agent. This causes A_3 to return **false** to A_1, which also returns **false** and terminates **MarkNode**.

- When **MarkedAll** is activated in A_1, it returns **false** immediately. **TestAcyclicity** is terminated with the conclusion that the DAG union is **cyclic**.

As the second example, consider the DAG union in Figure 9.15(a).

- Suppose A_1 is selected as A_*. Figure 9.18(a) shows the DAG union after **PreProcess** has been called on A_1.

- When **MarkNode** is first called on A_1, it calls **CollectFamilyInfo(a)** on itself and then **CollectFamilyInfo(b)** as in the previous example.

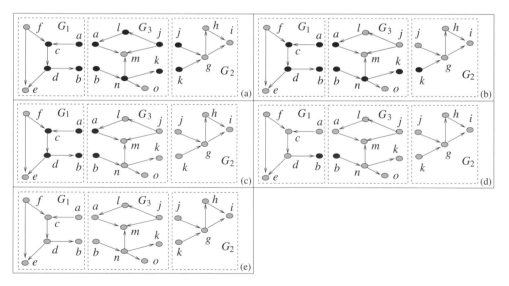

Figure 9.18: Illustration of **TestAcyclicity** in an acyclic DAG union.

- Agent A_1 calls **MarkNode** in Agent A_3, and calls on itself **CollectFamilyInfo(j)** and is then propagated to A_2. Agent A_2 returns a triplet $(j, 0, 0)$ to A_3. Subsequently, A_3 generates the final triplet $(j, 0, 1)$ and terminates **CollectFamilyInfo(j)**.

  Because $p = 0$, A_3 calls **DistributeMark(j)** on itself. It marks j and l and then calls **DistributeMark(j)** in A_2, which marks j as well. Figure 9.18(b) shows the resultant DAG union.

  Agent A_3 then calls on itself **CollectFamilyInfo(k)**, which eventually terminates similarly as **CollectFamilyInfo(j)**. Figure 9.18(c) shows the resultant DAG union.

- Agent A_3 calls **MarkNode** in A_2, which returns **false**. Because **DistributeMark** was called on A_3, it returns **true** to A_1, which returns **true** and terminates the first call of **MarkNode** in **TestAcyclicity**.

- When **MarkNode** is called on A_1 the second time, A_1 calls on itself **CollectFamily-Info(a)**, which is then propagated to A_3. Agent A_3 returns a triplet $(a, 0, 0)$ to A_1. Subsequently, A_1 generates the final triplet $(a, 0, 1)$ and terminates **CollectFamilyInfo(a)**.

  Because $p = 0$, A_1 calls **DistributeMark(a)** on itself. This causes the marking of a, c, d by A_1 and a by A_3. Figure 9.18(d) shows the resultant DAG union.

  Agent A_3 then calls on itself **CollectFamilyInfo(b)**. It eventually terminates similarly as **CollectFamilyInfo(a)** with b marked in both A_1 and A_3. Figure 9.18(e) shows the resultant DAG union.

- Agent A_1 calls **MarkNode** in A_3, which then calls **MarkNode** in A_2. Agent A_2 has no additional adjacent agent and returns **false**, which causes A_3 to return **false** to A_1. Because **DistributeMark** was called on A_1, it returns **true** and terminates the second call of **MarkNode** in **TestAcyclicity**.

- When **MarkNode** is called on A_1 the third time within **TestAcyclicity**, it propagates **MarkNode** to A_3 and then to A_1. Eventually, **false** is returned.

- When **MarkedAll** is called on A_1, it is propagated to A_3 and A_2. Eventually, **true** is returned.

- **TestAcyclicity** is now terminated with the conclusion that the DAG union is **acyclic**.

9.5.6 Correctness and Complexity

In the following theorem, we show that multiagent cooperative verification is achieved correctly by **TestAcyclicity**.

Theorem 9.5 *TestAcyclicity correctly determines the acyclicity of a hypertree DAG union.*

Proof: According to Corollary 9.4, it is sufficient to alternate marking of private roots and leaves and public roots and leaves recursively. **PreProcess** does the first round of recursive marking of private roots and leaves and repeated **MarkNode** performs the subsequent recursive and alternate marking. Each **MarkNode** identifies public roots and leaves by **CollectFamilyInfo** (either $p = 0$ or $c = 0$) and then marks them as well as new private roots and leaves by **DistributeMark**. By Corollary 9.4, **MarkNode** will not return **false** until all roots and leaves are marked. **MarkedAll** tests if the DAG union is empty and determines the acyclicity correctly.   □

Next, we establish that multiagent cooperative verification by **TestAcyclicity** is efficient. We use the following notations in the analysis:

- m: the maximum number of nodes in a local DAG.
- t: the maximum cardinality of a node adjacency in a local DAG.
- k: the maximum number of nodes in an agent interface.
- s: the maximum number of agents that share a given variable.
- n: the total number of agents in the hypertree DAG union.

To mark all private nodes recursively in a local DAG, $O(t\ m)$ nodes need to be checked. Hence, $O(n\ t\ m)$ nodes are checked during **PreProcess**.

An order of $O(s)$ agents are involved in each **CollectFamilyInfo** to test a single public node. To mark a public root or leaf, $O(s)$ agents are involved in the marking, and each of them also checks $O(t\ m)$ private nodes locally. Hence, the complexity of **CollectFamilyInfo** and **DistributeMark** for each marked public node is

$$O(s\ t\ m).$$

The complexity of each **MarkNode** called during **TestAcyclicity** is then $O(n\ k\ s\ t\ m)$ because **MarkNode** may be propagated to each agent and each may process up to k public nodes. Because at least one public node will be marked for each call of **MarkNode**, it will be called $O(n\ k)$ times. Hence, the complexity of all **MarkNode** calls is

$$O(n^2\ k^2\ s\ t\ m).$$

An order of $O(n\ m)$ nodes are checked during **MarkedAll**. Therefore, the time complexity of **TestAcyclicity** is

$$O(n^2\ k^2\ s\ t\ m).$$

This analysis shows implies that the multiagent cooperative verification is efficient.

To build a large system from components, two generic approaches towards correctness can be identified. One approach imposes more restrictions on the components to increase the chance that they integrate into a correct overall system. The other approach imposes less restrictions on the components but relies on system-wide tests to verify the system correctness. For example, in operating systems design (Silberschatz and Galvin [65]), *deadlock prevention* takes the first approach and *deadlock detection* takes the second.

The two approaches can both be applied to ensure the acyclicity of a hypertree DAG union. In OOBNs – i.e. (Koller and Pfeffer [35]), it is required that each network segment (similar to a local DAG) satisfy certain topological constraints so that the OOBN is always acyclic. For example, when two network segments share variables, the direction of all arcs points from one segment to the other. As a consequence, there is no need to verify acyclicity.

In this book, we do not impose such restrictions. When two local DAGs are adjacent on the hypertree, it is allowed to have a directed path going one way and another directed path in the opposite direction. For example, in Figure 9.15(a), a directed path (j, l, a, c, d, e) goes from G_3 into G_1 and another directed path (d, b, n, k) from G_1 into G_3.

As another example, consider the local DAGs in Figures 6.29 and 6.30 for the virtual digital components U_2 and U_3 (Figures 6.16 and 6.17). We see a directed path

$$(i_1, x_0, u_0, w_0, q_1)$$

going from G_2 into G_3, and another directed path

$$(p_1, n_1, y_0, u_0, w_0, z_4, n_0)$$

from G_3 into G_2. Such oppositely directed paths will be disallowed if the one-way-arcs restriction is imposed. However, they arise in practice. Using **TestAcyclicity**, we are assured that they do not cause cyclicity.

9.6 Verification of Agent Interface

9.6.1 The Issue of d-sepset Verification

According to Definition 6.15, each agent interface in an MSBN should be a d-sepset (Definition 6.11). An agent interface is a d-sepset if every public node in the interface is a d-sepnode (Definition 6.11). However, whether a public node in an interface I is a d-sepnode cannot be determined by the pair of local DAGs interfaced with I. According to Definition 6.11, given a hypertree DAG union G, whether a public node x in G is a d-sepnode depends on whether there exists a local DAG that contains all parents $\pi(x)$ of x in G. Any local DAG that shares x may potentially contain some parent nodes of x. Some parent nodes of x are public, but others are private. Because it is desirable not to disclose the details of the parentship for agent privacy, we cannot send the parents of x in each agent to a single agent for d-sepnode verification. Hence, cooperation among all agents whose subdomains contain parents of a public node x is required in order to verify whether x is a d-sepnode. Next we develop distributed algorithms that verify the d-sepset condition in a hypertree DAG union without violating agent privacy. We follow the naming convention used in previous sections on *caller*, A_c, A_0, A_1, . . ., A_k and A_*.

We organize agents' verification activities around the verification of the d-sepnode condition. If any public node x is found to be a non-d-sepnode, then any agent interface that contains x is not a d-sepset, and the hypertree DAG union has violated the d-sepset condition and is not a hypertree MSDAG. If no such node can be found, then every agent interface is a d-sepset and the d-sepset condition holds in the hypertree DAG union.

To verify the d-sepnode condition, we may use Definition 6.11 directly. However, that would require collecting parents of each public node from each agent and checking whether one of the agents contains them all. As mentioned previously, this violates agent privacy. In Section 9.6.2, we derive alternative conditions that can be used to verify the d-sepnode condition while respecting the agent privacy.

9.6.2 Checking Private Parents for Invalid Interface

The proposition below derives a *necessary* condition of a d-sepnode. It can be used to detect if a hypertree DAG union violates the d-sepset condition.

Proposition 9.6 *Let a public node x in a hypertree DAG union G be a d-sepnode. Then no more than one local DAG of G contains private parent nodes of x.*

Proof: We prove by contradiction. Assume that two or more local DAGs contain private parent nodes of x. Let y be a private parent of x contained in a local DAG G_i and z be a private parent of x contained in G_j $(i \neq j)$. Then there cannot be any one local DAG that contains both y and z. Hence, no local DAG contains all parents of x, and x is not a d-sepnode by Definition 6.11, which is a contradiction.   □

Proposition 9.6 requires only the information of whether an agent has private parents of a public node x. The proposition does not require the information on the number of parents nor on what they are. Hence, it admits agent privacy. Once the number of agents who have private parents of x are collected, the simple count is sufficient to determine if the hypertree DAG union has violated the d-sepset condition. Note, however, that Proposition 9.6 is not sufficient to qualify x as a d-sepnode and hence is not sufficient to qualify a hypertree DAG union as satisfying the d-sepset condition.

Given a public node x of agent A_0, the following **CollectPrivateParentInfo** determines how many agents on the subtree rooted at A_0 have private parent nodes

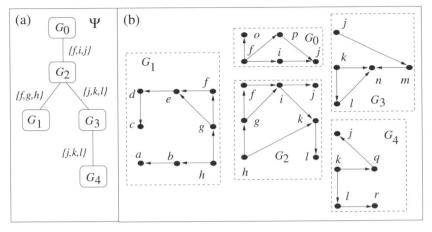

Figure 9.19: A hypertree DAG union with the hypertree in (a) and local DAGs in (b).

of x. According to Proposition 9.6, if more than one agent is found to have private parent nodes of x, then x is *not* a d-sepnode.

Algorithm 9.9 (CollectPrivateParentInfo) *When CollectPrivateParentInfo(x) is called on an agent A_0, it does the following:*

1. *Agent A_0 checks if x has private parents in G_0. It sets counter$_0$ = 1 if the result is positive and counter$_0$ = 0 otherwise.*
2. *If A_0 does not share x with any agent A_i ($i = 1, \ldots, k$), then A_0 returns counter$_0$ to caller. Otherwise, it continues.*
3. *For each agent A_i ($i = 1, \ldots, k$) that shares x, A_0 calls CollectPrivateParentInfo(x) in A_i, and when A_i returns counter$_i$, A_0 adds counter$_i$ to counter$_0$.*
4. *Agent A_0 returns counter$_0$ to caller.*

As an example, consider the hypertree DAG union in Figure 9.19. Suppose that **CollectPrivateParentInfo(j)** is called on A_0.

- Agent A_0 sets **counter$_0$** = 1 because G_0 contains a private parent p of j. Because A_0 shares j with the adjacent agent A_2, it calls A_2 to perform **CollectPrivateParentInfo(j)**.
- Agent A_2 has no private parent of j and sets **counter$_2$** = 0. Because it shares j with A_3, it calls A_3 to perform **CollectPrivateParentInfo(j)**.
- Agent A_3 has no private parent of j and sets **counter$_3$** = 0. It shares j with A_4 and hence calls A_4 to perform **CollectPrivateParentInfo(j)**.
- Agent A_4 has a private parent q of j and sets **counter$_4$** = 1. It has no adjacent agents except the caller A_3. Therefore, A_4 returns **counter$_4$** = 1 to A_3.
- Upon receipt of **counter$_4$** = 1, A_3 updates **counter$_3$** = 1. It then returns **counter$_3$** = 1 to A_2.

- The receipt of $counter_3 = 1$ causes A_2 to update $counter_2$ and returns $counter_2 = 1$ to A_0.
- Agent A_0 receives $counter_2 = 1$ and updates $counter_0 = 2$. It returns $counter_0 = 2$, which signifies that j is not a d-sepnode.

The following **FindNonDsepnodeByPrivateParent** algorithm is used by an agent A_0 to determine if each public node located on the subtree rooted at A_0 that is not shared by A_c violates the d-sepnode condition. The task is accomplished by calling **CollectPrivateParentInfo**.

Algorithm 9.10 (FindNonDsepnodeByPrivateParent) *When FindNonDsepnodeByPrivateParent is called on an agent A_0, it does the following:*

1. *Agent A_0 returns **false** if it has no adjacent agent except A_c. Otherwise, it continues.*
2. *For each public node x in G_0 such that x is not shared by A_c, A_0 calls on itself* **CollectPrivateParentInfo(x)**. *When **counter** is returned, A_0 returns **true** if **counter** exceeds 1. Otherwise, it saves $count(x) = counter$ and continues.*
3. *For each agent A_i ($i = 1, \ldots, k$), A_0 calls **FindNonDsepnodeByPrivateParent** in A_i. If A_i returns **true**, then A_0 returns **true**.*
4. *Agent A_0 returns **false**.*

We illustrate the execution of **FindNonDsepnodeByPrivateParent** using the hypertree DAG union in Figure 9.19. Suppose that **FindNonDsepnodeByPrivateParent** is called on A_0.

- Agent A_0 calls **CollectPrivateParentInfo(f)** on itself. It sets $counter_0 = 0$ and calls A_2 to perform **CollectPrivateParentInfo(f)**. Agent A_2 sets $counter_2 = 0$ and calls A_1 to perform **CollectPrivateParentInfo(f)**. When A_1 finishes, it returns $counter_1 = 0$ to A_2, which in turn returns $counter_2 = 0$ to A_0. Agent A_0 finishes **CollectPrivateParentInfo(f)** by returning $counter_0 = 0$. A record $count(f) = 0$ is made in A_0.
- Next, A_0 calls **CollectPrivateParentInfo(i)** on itself. It calls A_2 for **CollectPrivateParentInfo(i)** in due time. Because i has no private parent in G_0 and G_2, the process finishes with A_0 returning $counter_0 = 0$. A record $count(i) = 0$ is made in A_0.
- Agent A_0 calls **CollectPrivateParentInfo(j)** on itself. As detailed previously, eventually $counter_0 = 2$ is returned. Hence, A_0 returns **true** and terminates **FindNonDsepnodeByPrivateParent**.

Note that in **FindNonDsepnodeByPrivateParent**, when the returned counter has a value of 0 or 1, a record is made by the agent. This record is useful to differentiate shared nodes that have only public parents (when the value is 0) and those that have private parents (when the value is 1). We assume that this record survives the execution of **FindNonDsepnodeByPrivateParent**. Its usage is presented in Section 9.6.8.

We summarize the postcondition of **FindNonDsepnodeByPrivateParent** in the following proposition. The proof is trivial given Proposition 9.6.

Proposition 9.7 *Let G be a hypertree DAG union populated by a set of agents. Let* **FindNonDsepnodeByPrivateParent** *be called on any agent. If* **true** *is returned, then G does not satisfy the d-sepset condition. If* **false** *is returned, then for every public node x, no more than one local DAG of G contains private parent nodes of x.*

By Proposition 9.7, the verification is conclusive only when the execution of **FindNonDsepnodeByPrivateParent** returns **true**. Otherwise, further verification is necessary, as we discuss in the next section.

9.6.3 Processing Public Parents

When **FindNonDsepnodeByPrivateParent** returns **false** for a hypertree DAG union G, it is still possible that the d-sepset condition is not satisfied in G. Consider the example in Figure 9.20. The public nodes are w, x, y, z. No local DAG has any private parent of x or z. Only G_0 has a private parent of y, and only G_2 has a private parent of w. Hence **FindNonDsepnodeByPrivateParent** will return **false**. However, no single local DAG contains all parents of x: $\pi(x) = \{w, y\}$. Therefore, x is not a d-sepnode and none of the agent interfaces is a d-sepset.

By Proposition 9.7, if **FindNonDsepnodeByPrivateParent** returns **false**, then for each public node x either one local DAG or none contains private parents of x. To determine if G satisfies the d-sepset condition conclusively, one still needs to find

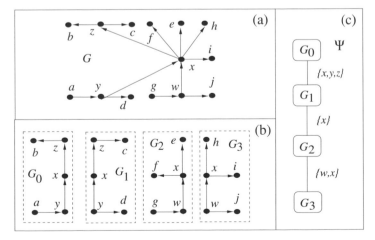

Figure 9.20: A hypertree DAG union G (a) with local DAGs in (b) and hypertree in (c).

out if there exists a local DAG that contains $\pi(x)$. It seems that we have not made much progress beyond brute force testing using Definition 6.11 directly. In fact, the remaining verification task is simpler than the original, for it is now sufficient to consider only the following two cases:

[Case 1] If one local DAG contains private parents of x, then it is the only candidate potentially to contain all parents of x. Hence, our verification effort can be more focused. The following proposition summarizes this idea. It is a direct result of Definition 6.11 and Proposition 9.6.

Corollary 9.8 *Let G be a hypertree DAG union. Let x be any public node such that a single local DAG G_i in G contains private parents of x. Then if G_i does not contain all parents of x in G, x is not a d-sepnode.*

In Section 9.6.8, we present algorithms to process this case.

[Case 2] If no local DAG contains any private parent of x, then all parents of x are public. Hence, verification can be performed by passing messages on these parents among agents without violating agent privacy. On the other hand, because any local DAG containing x may contain $\pi(x)$, there does not appear to be a single local DAG on which we can safely focus the verification processing.

To process this case, we analyze how elements in $\pi(x)$ may be distributed on the hypertree. Because local DAGs are organized into a hypertree, for each pair of local DAGs that contains some parents of x, every local DAG on the path between them should contain these parents. Furthermore, suppose that there exists a local DAG G_i containing $\pi(x)$. Let ρ be a directed path from G_i to any local DAG G_k containing some parents of x. Then along ρ, the parent nodes of x contained in each local DAG must be nonincreasing. That is, if one traverses from G_i along ρ, it is impossible to encounter a new parent node of x that has not been seen before.

On the other hand, as one traverses a path from a local DAG G_a to G_c both containing some parents of x, if a parent y of x is contained in G_a but not in G_c whereas another parent z of x is contained in G_c but not in G_a, and no local DAGs in between contain both y and z, then one can be certain that there cannot be any local DAG containing $\pi(x)$. This is what happens in the example of Figure 9.20. As one traverses the hypertree from G_0 to G_3, the parent y of x contained in G_0 and G_1 disappears in G_2 and G_3 while a new parent node w of x emerges. We explore these observations systematically in the next section.

9.6.4 Cooperative d-sepnode Testing in a Hyperchain

Consider first the situation in which no local DAG contains any private parent of x and all local DAGs containing either x or some public parents of x form a hyperchain

$\langle G_0, G_1, \ldots, G_m \rangle$. When $m = 1$, because none of G_0 and G_1 has private parents of x, clearly x is a d-sepnode. Hence, we only need to consider $m \geq 2$. In Definition 9.9 below we define the *public parent sequence* of x to describe the distribution of $\pi(x)$ on the hyperchain. The definition uses heavily set comparisons. Note that when two sets X and Y are compared, there are four possible outcomes: $X = Y$, $X \subset Y$, $X \supset Y$, or X and Y being incomparable (i.e., $X \nsubseteq Y$ and $X \nsupseteq Y$). We will use $X \bowtie Y$ to denote X and Y as incomparable. The symbol $\bowtie$ reads "bowtie."

Definition 9.9 *Let* $\langle G_0, G_1, \ldots, G_m \rangle$ $(m \geq 2)$ *be a hyperchain of local DAGs, where x is a public node, each G_i contains either x or some parents of x, and all parents of x are public. Denote the parents of x that G_i $(0 < i < m)$ shares with G_{i-1} and G_{i+1} by $\pi_i^-(x)$ and $\pi_i^+(x)$, respectively. Denote the parents of x that G_m shares with G_{m-1} by $\pi_m^-(x)$. Then the sequence*

$$(\pi_1^-(x), \pi_2^-(x), \ldots, \pi_m^-(x))$$

*is the **public parent sequence** of x on the hyperchain. The sequence is classified into the following types, where $0 < i < m$:*

Identical *For each i, $\pi_i^-(x) = \pi_i^+(x)$.*
Increasing *For each i, $\pi_i^-(x) \subseteq \pi_i^+(x)$, and there exists i such that $\pi_i^-(x) \subset \pi_i^+(x)$.*
Decreasing *For each i, $\pi_i^-(x) \supseteq \pi_i^+(x)$, and there exists i such that $\pi_i^-(x) \supset \pi_i^+(x)$.*
Concave *One of the following holds:*
  1. *For $m \geq 3$, there exists i such that the subsequence $(\pi_1^-(x), \ldots, \pi_i^-(x))$ is increasing and the subsequence $(\pi_i^-(x), \ldots, \pi_m^-(x))$, is decreasing.*
  2. *There exists i such that $\pi_i^-(x) \bowtie \pi_{i+1}^-(x)$; the preceding subsequence $(\pi_1^-(x), \ldots, \pi_i^-(x))$ is trivial $(i = 1)$, increasing, or identical; and the trailing subsequence $(\pi_{i+1}^-(x), \ldots, \pi_m^-(x))$ is trivial $(i = m - 1)$, decreasing, or identical.*
Wave *One of the following holds:*
  1. *There exists i such that $\pi_i^-(x) \supset \pi_i^+(x)$ and $j > i$ such that either $\pi_j^-(x) \subset \pi_j^+(x)$ or $\pi_j^-(x) \bowtie \pi_j^+(x)$.*
  2. *There exists i such that $\pi_i^-(x) \bowtie \pi_i^+(x)$ and $j > i$ such that either $\pi_j^-(x) \subset \pi_j^+(x)$ or $\pi_j^-(x) \bowtie \pi_j^+(x)$.*

Figure 9.21 illustrates the first three sequence types, where only x and its parents are shown explicitly in each agent interface. The identical sequence is illustrated in (a). Each G_i contains $\pi(x)$, and hence x is a d-sepnode. The increasing sequence is exemplified in (b). Because G_m contains $\pi(x)$, x is a d-sepnode. The decreasing sequence is exemplified in (c). It is symmetric to the increasing sequence; G_0 contains $\pi(x)$ and x is a d-sepnode.

For the concave sequence, some parents of x appear in the middle of the hyperchain but do not appear on either end. Figure 9.22 illustrates the possible cases.

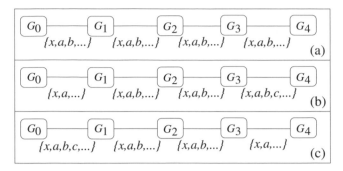

Figure 9.21: Public parent sequences: (a) An identical sequence; (b) An increasing sequence; (c) A decreasing sequence.

Case (1) is illustrated in (a), where b is contained in G_1, G_2, and G_3 but disappears in G_0 and G_4 and c is contained in G_2 and G_3 but disappears in G_0, G_1, and G_4. The increasing subsequence and the decreasing subsequence share $\pi_3^-(x)$. At least two local DAGs in the middle of the hyperchain will contain $\pi(x)$ (e.g., G_2 and G_3 in (a)), and hence x is a d-sepnode. The remainder of Figure 9.22 illustrates case (2). In (b), G_1 contains both b and c, G_0 contains only b, and G_2 contains only c. In other words, $\pi_1^-(x)$ and $\pi_2^-(x)$ are incomparable. Note that the subsequence $(\pi_1^-(x), \pi_2^-(x))$ is preceded and trailed by trivial subsequences. Clearly, G_1 contains $\pi(x)$ and x is a d-sepnode. In (c), the subsequence $(\pi_1^-(x), \pi_2^-(x))$ is incomparable and is trailed by a decreasing subsequence. Graph G_1 contains $\pi(x)$,

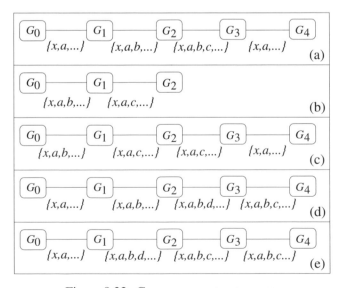

Figure 9.22: Concave parent sequences.

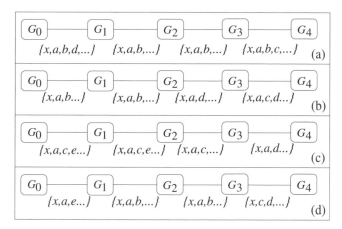

Figure 9.23: Wave parent sequences.

and x is a d-sepnode. In (d), the subsequence $(\pi_1^-(x), \ldots, \pi_3^-(x))$ is increasing with $\pi_3^-(x)$ and $\pi_3^+(x)$ incomparable. Graph G_3 contains $\pi(x)$ and x is a d-sepnode. In (e), the increasing subsequence ends at $\pi_2^-(x)$, and the decreasing subsequence starts at $\pi_3^-(x)$ with $\pi_2^-(x)$ and $\pi_3^-(x)$ incomparable. Because G_2 contains $\pi(x)$, x is a d-sepnode.

Public parent sequences of the identical, increasing, decreasing, and concave type represent all possible parent sequences, where $\pi(x)$ can be contained in one local DAG and hence cover all cases in which x is a d-sepnode. We will formally establish this shortly. Definition 9.9 lumps all other public parent sequences together under the wave type. Figure 9.23 illustrates several possible cases.

In (a), a parent d of x appears at one end of the hyperchain, another parent c appears at the other end, and they disappear in the middle of the hyperchain. In other words, we have $\pi_1^-(x) \supset \pi_1^+(x)$ and $\pi_3^-(x) \subset \pi_3^+(x)$. No local DAG contains all parents of x, and hence x is not a d-sepnode. In (b), we have $\pi_2^-(x) \bowtie \pi_2^+(x)$ and $\pi_3^-(x) \subset \pi_3^+(x)$. In (c), $\pi_2^-(x) \supset \pi_2^+(x)$ and $\pi_3^-(x) \bowtie \pi_3^+(x)$ hold. In (d), we have $\pi_1^-(x) \bowtie \pi_1^+(x)$ and $\pi_3^-(x) \bowtie \pi_3^+(x)$. In all these cases, no local DAG contains $\pi(x)$, and x is not a d-sepnode.

In the preceding examples, each local DAG contains both x and some parents of x. This may not always be the case. In general, some local DAGs may contain x but none of the elements of $\pi(x)$, whereas other local DAGs may contain some parents of x but not x itself. Figure 9.24 shows a hyperchain $\langle G_0, G_1, G_2, G_3, G_4 \rangle$, where x is a public node contained in local DAGs G_1, G_2, G_3, and G_4. The set of parents $\pi(x)$ of x in the DAG union is $\{w, y\}$, and each element of $\pi(x)$ is public. Three local DAGs contain both x and some parents of x: G_1 contains both x and y, and G_2 and G_3 contain both x and w. One local DAG, G_4, contains x but no parent of x. One local DAG, G_0, contains a parent y of x but not x. The public parent sequence

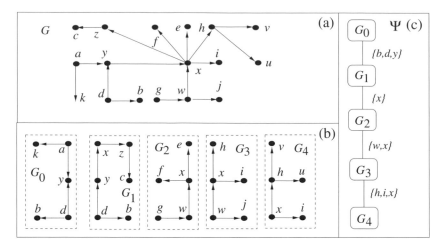

Figure 9.24: A hyperchain DAG union G with local DAGs in (b) and hyperchain in (c). Graph G_0 contains a public parent of a public node x without containing x itself.

of x on the hyperchain is

$$(\pi_1^-(x), \pi_2^-(x), \pi_3^-(x), \pi_3^+(x)) = (\{y\}, \{\}, \{w\}, \{\}).$$

Theorem 9.10 shows that the identical, increasing, decreasing, and concave types represent all possible parent sequences, where $\pi(x)$ can be contained in one local DAG.

Theorem 9.10 *Let x be a public node in a hyperchain $\langle G_0, G_1, \ldots, G_m \rangle$ of local DAGs with $\pi(x)$ being the parents of x in all DAGs, where no parent of x is private and each local DAG contains either x or some parents of x.*

Then there exists one local DAG that contains $\pi(x)$ if and only if the public parent sequence of x on the hyperchain is identical, increasing, decreasing, or concave.

Proof: [Sufficiency] If the sequence type is identical, then every local DAG contains $\pi(x)$. If the type is increasing, then at least G_m contains $\pi(x)$. If the type is decreasing, then at least G_0 contains $\pi(x)$. If the type is concave, for Case (1) (see Definition 9.9), both G_i and G_{i-1} contain $\pi(x)$. For case (2), G_i contains $\pi(x)$.

[Necessity] Suppose that there exists a local DAG that contains $\pi(x)$. We show that the parent sequence of x is identical, increasing, decreasing, or concave.

If every local DAG contains $\pi(x)$, then $\pi_j^-(x) = \pi_j^+(x)$ for each j and the sequence is identical. Otherwise, if G_0 contains $\pi(x)$, then $\pi_j^-(x) \supseteq \pi_j^+(x)$ for each j and the sequence is decreasing. Otherwise, if G_m contains $\pi(x)$, then $\pi_j^-(x) \subseteq \pi_j^+(x)$ for each j and the sequence is increasing.

Otherwise, if both G_i and G_{i-1} contain $\pi(x)$ for some i ($2 \leq i \leq m-1$), then $\pi_j^-(x) \subseteq \pi_j^+(x)$ for each $j \leq i-1$ and the subsequence $(\pi_1^-(x), \ldots, \pi_i^-(x))$, is increasing, and $\pi_j^-(x) \supseteq \pi_j^+(x)$ for each $j \geq i$, and the subsequence $(\pi_i^-(x), \ldots, \pi_m^-(x))$ is decreasing. The entire parent sequence falls under concave type Case (1).

Otherwise, if only one local DAG G_i contains $\pi(x)$, then

$$\pi_i^-(x) \subset \pi(x) \quad \text{and} \quad \pi_i^+(x) \subset \pi(x).$$

We show $\pi_i^-(x) \bowtie \pi_i^+(x)$ by contradiction. If they are comparable, then

$$\text{either } \pi_j^-(x) \subseteq \pi_j^+(x) \quad \text{or} \quad \pi_j^-(x) \supset \pi_j^+(x).$$

We have

$$\pi_j^-(x) \subseteq \pi_j^+(x) \subset \pi(x) \quad \text{or} \quad \pi(x) \supset \pi_j^-(x) \supset \pi_j^+(x),$$

which implies that $\pi(x)$ contains a private parent of x: a contradiction. Furthermore, the subsequence $(\pi_1^-(x), \ldots, \pi_i^-(x))$ must be trivial, increasing, or identical, and the subsequence $(\pi_{i+1}^-(x), \ldots, \pi_m^-(x))$ must be trivial, decreasing, or identical. Hence, the entire parent sequence falls under concave type Case (2). $\qquad\square$

Theorem 9.11 shows that the wave public parent sequences characterize all possible parent sequences when there is no local DAG containing $\pi(x)$.

Theorem 9.11 *Let x be a public node in a hyperchain $\langle G_0, G_1, \ldots, G_m \rangle$ of local DAGs with $\pi(x)$ being the parents of x in all DAGs, where no parent of x is private and each local DAG contains either x or some parents of x.*

Then there exists no local DAG that contains $\pi(x)$ if and only if the public parent sequence of x on the hyperchain is a wave.

Proof: [Sufficiency] Suppose that the sequence is of the wave type. For the sequence is of the wave type Case (1) in Definition 9.9, we have $\pi_i^-(x) \supset \pi_i^+(x)$. It implies that G_{i-1} and G_i contain a parent, say y, of x that is not contained in G_{i+1}. It cannot be contained in any G_k where $k > i+1$ owing to the hyperchain. If $\pi_j^-(x) \subset \pi_j^+(x)$ holds, then G_{j+1} and G_j contain a parent, say z, of x that is not contained in G_{j-1}. It cannot be contained in any G_k, where $k < j-1$. In summary, only local DAGs $G_0, \ldots, G_i$ may contain y (not necessarily all of them contain y), and only $G_j, \ldots, G_m$ may contain z. Because $i < j$, no local DAG contains both y and z.

If $\pi_j^-(x) \bowtie \pi_j^+(x)$, it implies that G_{j+1} and G_j contain a parent, say z, of x that is not contained in G_{j-1}, and G_{j-1} and G_j contains a parent, say w, of x that is not contained in G_{j+1}. Because the same condition as above holds, no local DAG contains both y and z. For wave type case (2), the same conclusion can be drawn.

[Necessity] Suppose that no local DAG contains $\pi(x)$. Then a pair of local DAGs G_i and G_j $(i < j)$ exist such that the following hold:

1. The DAG G_i contains a parent, say y, of x that is not contained in G_j, and G_i is the closest such local DAG to G_j on the hyperchain.
2. The DAG G_j contains a parent, say z, of x that is not contained in G_i, and G_j is the closest such local DAG to G_i on the hyperchain.
3. No other local DAGs contain both y and z.

Clearly, we have either $\pi_i^-(x) \supset \pi_i^+(x)$ or $\pi_i^-(x) \bowtie \pi_i^+(x)$, and either $\pi_j^-(x) \subset \pi_j^+(x)$ or $\pi_j^-(x) \bowtie \pi_j^+(x)$. Hence, the sequence is of the wave type. $\qquad\square$

The following proposition establishes that the five public parent sequence types defined so far cover all possibilities.

Proposition 9.12 *Let $\langle G_0, G_1, \ldots, G_m \rangle$ $(m \geq 2)$ be a hyperchain of local DAGs, where x is a public node, each G_i contains either x or some parents of x, and all parents of x are public.*

Then the five public parent sequence types defined in Definition 9.9 are exhaustive.

Proof: Because either there exists a local DAG that contains $\pi(x)$ or such a local DAG does not exist. By Theorem 9.10, the former case is equivalent to the identical, increasing, decreasing, or concave types. By Theorem 9.11, the latter case is equivalent to the wave type. Hence, the five types are exhaustive. $\qquad\square$

The following corollary summarizes the relation between the type of a public parent sequence and the nature of a public node x. Given Theorems 9.10 and 9.11, its proof is straightforward.

Corollary 9.13 *Let x be a public node in a hyperchain DAG union $\langle G_0, G_1, \ldots, G_m \rangle$ $(m \geq 2)$, where no local DAG contains any private parent of x and each local DAG contains either x or some public parents of x.*

Then x is a d-sepnode if the public parent sequence of x on the hyperchain is identical, increasing, decreasing, or concave, and x is a non-d-sepnode if the sequence is wave.

To identify the sequence type by cooperation, let agents on the hyperchain pass messages from one end to the other, say, from G_m to G_0. Each agent A_i passes a message to A_{i-1} formulated based on the message that A_i receives from A_{i+1} as well as on the result of comparison between $\pi_i^-(x)$ and $\pi_i^+(x)$. Note that A_{i+1} is undefined for A_m. A message is represented using the following symbols:

IDEN (for identical), INC (for increasing), DEC (for decreasing), CONC (for concave), or WAVE (for wave). Messages are formulated according to the following algorithm:

Algorithm 9.11 (CollectPublicParentInfoOnChainBySymbol) *If A_{i+1} is undefined, agent A_i passes IDEN to A_{i-1}. Otherwise, A_i receives a message from A_{i+1}, compares $\pi_i^-(x)$ with $\pi_i^+(x)$, and sends its own message according to one of the following cases:*

1. *The message received is IDEN:*
   If $\pi_i^-(x) = \pi_i^+(x)$, A_i passes IDEN to A_{i-1}.
   If $\pi_i^-(x) \supset \pi_i^+(x)$, A_i passes DEC to A_{i-1}.
   If $\pi_i^-(x) \subset \pi_i^+(x)$, A_i passes INC to A_{i-1}.
   Otherwise, A_i passes CONC to A_{i-1}.
2. *The message received is DEC:*
   If $\pi_i^-(x) \supseteq \pi_i^+(x)$, A_i passes DEC to A_{i-1}.
   Otherwise, A_i passes CONC to A_{i-1}.
3. *The message received is INC:*
   If $\pi_i^-(x) \subseteq \pi_i^+(x)$, A_i passes INC to A_{i-1}.
   Otherwise, A_i passes WAVE to A_{i-1}.
4. *The message received is CONC:*
   If $\pi_i^-(x) \subseteq \pi_i^+(x)$, A_i passes CONC to A_{i-1}.
   Otherwise, A_i passes WAVE to A_{i-1}.
5. *The message received is WAVE: A_i passes WAVE to A_{i-1}.*

As examples, consider the sequences in Figure 9.21.

- In (a), A_4 sends IDEN to A_3, which is passed along by each agent until A_0 receives it.
- In (b), A_4 sends IDEN to A_3, which in turn sends INC to A_2. The message INC will be passed all the way to A_0 because $\pi_i^-(x) \subseteq \pi_i^+(x)$ holds for $i = 2$ and $i = 1$.
- In (c), A_3 receives IDEN and sends DEC to A_2. The message DEC will be passed all the way to A_0.

Next, consider the concave sequences in Figure 9.22.

- In (a), A_3 sends DEC to A_2, which in turn sends CONC to A_1. The message is then passed to A_0.
- Agent A_1 sends CONC to A_0 in (b).
- In (c), A_3 sends DEC to A_2, which passes it to A_1. Agent A_1 sends CONC to A_0 based on its comparison.
- In (d), A_3 sends CONC to A_2, which is passed all the way to A_0.
- In (e), A_3 sends IDEN to A_2, which in turn sends CONC to A_1. The message is then passed to A_0.

Finally, consider the wave sequences in Figure 9.23.

- In (a), A_3 sends INC to A_2, which passes it to A_1. Because $\pi_1^-(x) \supset \pi_1^+(x)$, A_1 sends WAVE to A_0.
- In (b), A_3 sends INC to A_2, which in turn sends WAVE to A_1. Agent A_1 sends WAVE to A_0.
- In (c), A_3 sends CONC to A_2, which in turn sends WAVE to A_1. The message is then sent by A_1 to A_0.
- In (d), A_3 sends CONC to A_2, which is passed to A_1. Because $\pi_1^-(x) \bowtie \pi_1^+(x)$, A_1 sends WAVE to A_0.

Given a hyperchain $\langle G_0, G_1, \ldots, G_m \rangle$, according to **CollectPublicParentInfoOnChainBySymbol**, only the message IDEN sent by agent A_m is a message by default. All messages sent by other agents are consistent with Definition 9.9 and the intended interpretation of the message symbols. This has been demonstrated from the preceding discussion of examples in Figures 9.21 through 9.23. Hence, from the message that A_0 receives, the type of the public parent sequence can be determined. Whether x is a d-sepnode can in turn be concluded from the sequence type according to Corollary 9.13. For instance, if A_0 receives CONC, then it knows that the parent sequence is concave and x is a d-sepnode. If A_0 receives WAVE instead, then it knows that the parent sequence is the wave type and x is a non-d-sepnode. We will not establish these results formally here but will do so for a simplified version of **CollectPublicParentInfoOnChainBySymbol** in Section 9.6.5.

9.6.5 Consolidation of d-sepnode Testing Messages

In Section 9.6.4, we have developed a set of messages that can be used to determine the nature of a public node in a hyperchain of local DAGs. The set of messages consists of five distinct symbols: IDEN, DEC, INC, CONC, and WAVE. Because we only need to classify a public node x into either a d-sepnode and non-d-sepnode, the message symbols may be consolidated while maintaining the effectiveness of classification. We explore this possibility below with the goal of simplifying the d-sepnode testing through the consolidation.

We partition the five public parent sequence types into three groups: identical or decreasing, increasing or concave, and wave. We associate each group with a message coded using an integer, as in Table 9.1.

As was the case with **CollectPublicParentInfoOnChainBySymbol**, we will let agents on a hyperchain pass messages from A_m to A_0. Using the message code, we modify the algorithm **CollectPublicParentInfoOnChainBySymbol** into the following algorithm: **CollectPublicParentInfoOnChain**.

Table   9.1: *Message code for*
grouping public parent
sequence types

Type group	Code
decreasing or identical	−1
increasing or concave	1
wave	0

Algorithm 9.12 (CollectPublicParentInfoOnChain) *If A_{i+1} is undefined, agent A_i passes −1 to A_{i-1}. Otherwise, A_i receives a message from A_{i+1}, compares $\pi_i^-(x)$ with $\pi_i^+(x)$, and sends its own message according to one of the following cases:*

1. *The message received is −1:*
   If $\pi_i^-(x) \supseteq \pi_i^+(x)$, A_i passes −1 to A_{i-1}.
   Otherwise, A_i passes 1 to A_{i-1}.
2. *The message received is 1:*
   If $\pi_i^-(x) \subseteq \pi_i^+(x)$, A_i passes 1 to A_{i-1}.
   Otherwise, A_i passes 0 to A_{i-1}.
3. *The message received is 0: A_i passes 0 to A_{i-1}.*

The following proposition ensures that the outgoing message of each agent on the hyperchain is always defined. Its proof is straightforward.

Proposition 9.14 *Let x be a public node in a hyperchain DAG union $\langle G_0, G_1, \ldots, G_m \rangle$ ($m \geq 2$), where all parents of x are public and each local DAG contains either x or some public parents of x. Let the hyperchain be populated by a set of agents with one for each local DAG.*

*Then the outgoing message by each agent A_i according to **CollectPublicParentInfoOnChain** is well defined.*

We illustrate **CollectPublicParentInfoOnChain** with the examples in Figures 9.21 through 9.23. First, consider the parent sequences in Figure 9.21.

- In (a), −1 is sent from A_4 to A_3 and is passed along by each agent until A_0 receives it. Interpreting the message code, A_0 concludes that the parent sequence is either identical or decreasing. Because the actual sequence is identical, the conclusion is correct.
- In (b), A_3 receives −1 from A_4 and sends 1 to A_2. Afterwards, 1 is passed all the way to A_0, which determines that the sequence is either increasing (actual type) or concave.

- In (c), -1 is sent by each agent. The conclusion drawn by A_0 is to classify the type of sequence as either identical or decreasing (actual type).

The parent sequences in Figure 9.22 cause the following messages to be passed:

- In (a), A_3 receives -1 from A_4 and sends -1 to A_2. Agent A_2 sends 1 to A_1, which passes it to A_0. Agent A_0 then concludes that the sequence type is either increasing or concave, where concave is the actual type.
- In (b), A_1 sends 1 to A_0, which reaches the same conclusion as in (a).
- In (c), -1 is passed along from A_4 until A_1, which sends 1 to A_0.
- In (d), A_3 sends 1 to A_2, which is passed along to A_0.
- In (e), -1 is sent from A_4 to A_3 and then to A_2. Agent A_2 sends 1 to A_1, which is passed to A_0.

The wave parent sequences in Figure 9.23 cause the following messages to be passed:

- In (a), A_3 receives -1 from A_4 and sends 1 to A_2. Agent A_2 passes 1 to A_1, which in turn sends 0 to A_0. Agent A_0 then interprets the sequence type as a wave, which matches the actual type.
- In (b), A_3 receives -1 from A_4 and sends 1 to A_2. Agent A_2 sends 0 to A_1, which passes 0 to A_0.
- In (c), A_3 receives -1 and sends 1 to A_2. Agent A_2 sends 0 to A_1, which passes 0 to A_0.
- In (d), A_3 receives -1 and sends 1 to A_2. Agent A_2 passes 1 to A_1, which sends 0 to A_0.

Proposition 9.15 establishes that, after message passing according to **Collect-PublicParentInfoOnChain**, the type of parent sequence and the nature of public node x can be determined correctly. This proposition also covers the case of a trivial hyperchain ($m = 1$), where A_0 will receive -1 from A_1, and x is a d-sepnode.

Proposition 9.15 *Let x be a public node in a hyperchain DAG union $\langle G_0, G_1, \ldots, G_m \rangle$ ($m > 0$), where all parents of x are public and each local DAG contains either x or some public parents of x. Let the hyperchain be populated by agents with one for each local DAG.*

Then, after agents pass messages along the hyperchain from A_m to A_0 according to ***CollectPublicParentInfoOnChain****, the following hold:*

1. *The message code received by A_0 identifies the type of public parent sequence of x correctly.*
2. *If A_0 receives -1 or 1, then x is a d-sepnode.*
3. *If A_0 receives 0, then x is a non-d-sepnode.*

Proof: We only have to prove the first statement. Once it is proven, the next two statements follow according to Corollary 9.13. To prove the first statement, we show that whenever the sequence is of types identical or decreasing, A_0 receives -1; whenever the sequence is increasing or concave, A_0 receives 1; and whenever the sequence is the wave type, A_0 receives 0. Because the set of five sequence types is exhaustive by Proposition 9.12, the correctness of the first statement will be established.

1. We show that whenever the sequence is an identical or decreasing type, A_0 receives -1. Assume that the sequence is identical or decreasing. We claim that each agent will receive -1 and send out -1. Agent A_m will send -1 by default. We only need to show that if A_i receives -1, it will send -1. From Definition 9.9, A_i will find $\pi_i^-(x) \supseteq \pi_i^+(x)$. According to **CollectPublicParentInfoOnChain**, the outgoing message is -1.

2. We show that whenever the sequence is increasing, A_0 receives 1. By Definition 9.9, each A_i will find $\pi_i^-(x) \subseteq \pi_i^+(x)$ and at least one will find $\pi_i^-(x) \subset \pi_i^+(x)$. Let A_j be the first agent on the hyperchain that finds $\pi_j^-(x) \subset \pi_j^+(x)$. Because each A_i ($j < i < m$) will receive -1 and send -1, A_j will receive -1 and send 1 according to **CollectPublicParentInfoOnChain**. Each A_i ($0 < i < j$) will receive 1 and send 1 given that to $\pi_i^-(x) \subseteq \pi_i^+(x)$.

3. We show that whenever the sequence is concave, A_0 receives 1. For case (1) of the concave type in Definition 9.9, there exists an agent A_i such that the parent sequence on the (sub)hyperchain $\langle G_{i-1}, \ldots, G_m \rangle$ is decreasing. From the foregoing analysis, A_{i-1} will receive -1 from A_i. The (sub)hyperchain $\langle G_0, \ldots, G_{i-1} \rangle$ is increasing, and by Definition 9.9 each agent A_j ($0 < j < i - 1$) will find $\pi_j^-(x) \subseteq \pi_j^+(x)$ with at least one that finds $\pi_j^-(x) \subset \pi_j^+(x)$. Let A_k be the first agent on the (sub)hyperchain that finds $\pi_k^-(x) \subset \pi_k^+(x)$. Each A_j ($k < j < i - 1$) will receive -1 and send -1. Agent A_k will receive -1 and send 1 by **CollectPublicParentInfoOnChain**. Each agent A_j ($0 < j < k$) will receive 1 and send 1.

   For case (2) of the concave type in Definition 9.9, an agent A_i exists that finds $\pi_i^-(x) \bowtie \pi_i^+(x)$. If $i = m - 1$ (trivial trailing case), A_i will receive -1 by default. If the parent sequence on the (sub)hyperchain $\langle G_i, \ldots, G_m \rangle$ is decreasing or identical, A_i will also receive -1 by **CollectPublicParentInfoOnChain**. Because $\pi_i^-(x) \bowtie \pi_i^+(x)$, A_i will send 1. If $i = 1$ (trivial preceding case), the receiver from A_i is A_0. If the parent sequence on the (sub)hyperchain $\langle G_0, \ldots, G_i \rangle$ is increasing or identical, according to **CollectPublicParentInfoOnChain** A_{i-1} will pass 1 all the way to A_0.

4. We show whenever the sequence is the wave type that A_0 receives 0. By Definition 9.9, when the sequence is the wave type, there are at least two agents A_i and A_j ($i < j$) such that A_i finds either $\pi_i^-(x) \supset \pi_i^+(x)$ or $\pi_i^-(x) \bowtie \pi_i^+(x)$ and A_j finds either $\pi_j^-(x) \subset \pi_j^+(x)$ or $\pi_j^-(x) \bowtie \pi_j^+(x)$. We assume that A_i and A_j are two such agents that are closest to A_m on the hyperchain. This implies that each agent A_k ($j < k$) must find $\pi_k^-(x) \supseteq \pi_k^+(x)$. Hence, A_j receives -1 and sends out 1.

Because A_i is the closest agent to A_j, which finds either $\pi_i^-(x) \supset \pi_i^+(x)$ or $\pi_i^-(x) \bowtie \pi_i^+(x)$, each agent A_k ($i < k < j$) between A_i and A_j will find $\pi_k^-(x) \subseteq \pi_k^+(x)$. According to **CollectPublicParentInfoOnChain**, A_k will receive 1 and send out 1. As a result, A_i will receive 1 and send out 0, which will be passed along to A_0. $\qquad\qquad\square$

9.6.6 Cooperative d-sepnode Testing in a Hyperstar

Before moving on to the general hyperstar, we consider message passing by agents from both ends of a hyperchain toward an agent A_i in the middle. This can be viewed as a special case of a hyperstar with only two terminals. We refer to A_i as the *center agent*. Readers should keep in mind that when message passing is performed from both ends of the hyperchain, it should be understood as two separate executions of **CollectPublicParentInfoOnChain**. One is performed on the (sub)hyperchain

$$\langle G'_0, \ldots, G'_j, \ldots, G'_{m-i} \rangle, \quad \text{where } G'_j = G_{j+i},$$

and the other on the (sub)hyperchain

$$\langle G'_0, \ldots, G'_j, \ldots, G'_i \rangle, \quad \text{where } G'_j = G_{i-j}.$$

For example, if the hyperchain is $\langle G_0, \ldots, G_9 \rangle$ and the center agent is A_5, then the two executions are performed on $\langle G_5, \ldots, G_9 \rangle$, where messages propagate from A_9 to A_5, and on $\langle G_5, \ldots, G_0 \rangle$, where messages propagate from A_0 to A_5.

In Section 9.6.4, the only information that agent A_0 needs to process is the message received from A_1. Here, the center agent A_i has three pieces of information: two messages received from adjacent agents and a comparison between $\pi_i^-(x)$ and $\pi_i^+(x)$. The key to determine whether x is a d-sepnode is to detect whether its public parent sequence is the wave type. A wave sequence can be detected based on one message received by A_i only, or if not sufficient based on both messages received, or if still not sufficient based in addition on the comparison between $\pi_i^-(x)$ and $\pi_i^+(x)$. The following proposition establishes the correctness of this strategy. It uses the notation $X \nsubseteq Y$ and $X \nsupseteq Y$ for two sets X and Y, where $X \nsubseteq Y$ means either $X \supset Y$ or $X \bowtie Y$.

Proposition 9.16 *Let x be a public node in a hyperchain DAG union $\langle G_0, G_1, \ldots, G_m \rangle$ ($m > 1$), where all parents of x are public and each local DAG contains either x or some parents of x. Let the hyperchain be populated by agents with one for each local DAG.*

*After agents pass messages along the hyperchain from A_m and A_0 towards A_i ($0 < i < m$), according to **CollectPublicParentInfoOnChain**, the following hold:*

1. *If A_i receives 0 from one adjacent agent, then x is a non-d-sepnode.*
2. *Otherwise, if 1 is received by A_i from both adjacent agents, then x is a non-d-sepnode.*
3. *Otherwise, if 1 is received by A_i from A_{i+1} and $\pi_i^-(x) \nsubseteq \pi_i^+(x)$ or 1 is received from A_{i-1} and $\pi_i^-(x) \nsupseteq \pi_i^+(x)$, then x is a non-d-sepnode.*
4. *Otherwise, x is a d-sepnode.*

Proof: We show that if x is a non-d-sepnode, one of the first three cases will occur, and if x is a d-sepnode, none of the first three cases will occur.

Assume that x is a non-d-sepnode. According to Theorem 9.11, x is a non-d-sepnode if and only if its parent sequence is the wave type. By Definition 9.9, when the sequence is the wave type, there are at least two agents A_j and A_k $(j < k)$ such that A_j finds either $\pi_j^-(x) \supset \pi_j^+(x)$ or $\pi_j^-(x) \bowtie \pi_j^+(x)$ and A_k finds either $\pi_k^-(x) \subset \pi_k^+(x)$ or $\pi_k^-(x) \bowtie \pi_k^+(x)$. We assume that A_j and A_k are two such agents that are closest to A_m on the hyperchain. The position of A_j and A_k relative to A_i can be described by the following, excluding symmetric conditions:

Cond1 $i < j < k$: A_j and A_k are at the same side of A_i on the hyperchain.
Cond2 $j < i < k$: A_i is between A_j and A_k.
Cond3 $i = j < k$: one of A_j and A_k is the center agent A_i.

The parent sequence can be viewed as two subsequences each from the center of the original sequence to one terminal. In Cond1, because A_j and A_k both involve the subsequence from A_i to A_m, by Proposition 9.15 A_i will receive 0 as in case (1).

In Cond2, A_i will receive 1 from A_{i+1} as argued in the proof of Proposition 9.15. By symmetry, A_i will also receive 1 from A_{i-1}. This is case (2) of the proposition.

In Cond3, A_i will receive 1 from A_{i+1} and -1 from A_{i-1}. It will find $\pi_i^-(x) \nsubseteq \pi_i^+(x)$. This is case (3) of the proposition.

Next, we assume that x is a d-sepnode and show that none of the first three cases will occur. By Proposition 9.12, the parent sequence is identical, increasing, decreasing, or concave. We consider each individually.

When the sequence is identical, A_i will receive -1 from both adjacent agents. Hence, none of the first three cases will occur.

When the sequence is increasing, there exists at least one agent A_j that will find $\pi_j^-(x) \subset \pi_j^+(x)$. Let A_j be the closest such agent to A_m. Hence, each agent A_k $(j < k)$ will find $\pi_k^-(x) = \pi_k^+(x)$, and each agent A_k $(k < j)$ will find $\pi_k^-(x) \subseteq \pi_k^+(x)$. The position of A_j relative to A_i can be described by the following conditions:

Cond4 $i < j$: A_j and A_m are at the same side of A_i on the hyperchain.
Cond5 $j < i$: A_j and A_0 are at the same side of A_i.
Cond6 $i = j$: A_j is the center agent A_i.

In Cond4, A_j will receive -1 and send out 1, which will be passed to A_i by A_{i+1}. Because the subsequence between A_0 and A_i is identical or decreasing viewed from A_i to A_0, A_i will receive -1 from A_{i-1}. Hence, Cases (1) and (2) will not occur. Inasmuch as A_i will find $\pi_i^-(x) \subseteq \pi_i^+(x)$, Case (3) will not occur either.

In Cond5 and Cond6, A_i will receive -1 from A_{i+1} and -1 from A_{i-1} by the similar argument above. Hence, none of the first three cases will occur.

The decreasing sequence is symmetric to the increasing sequence, and we only have the concave sequence left to consider. By Definition 9.9, when the sequence is concave, a distinctive agent A_j exists. In case 1 of the concave type, the subsequence $(\pi_1^-(x), \dots, \pi_j^-(x))$ is increasing and $(\pi_j^-(x), \dots, \pi_m^-(x))$ is decreasing. In Case (2) of the concave type, $\pi_j^-(x) \bowtie \pi_j^+(x)$ holds. The position of A_j relative to A_i can be described by Cond4, Cond5, and Cond6.

Consider Case (1) of the concave-type sequences. In Cond4, A_j will receive -1 and the remaining analysis is the same as the increasing sequence case. Cond5 is symmetric to Cond4. In Cond6, A_i will receive -1 from both adjacent agents. For Case (2) of the concave type, the analysis is similar. $\square$

We now consider a general hyperstar with two or more terminals. The structure from the center of the hyperstar to each terminal is a hyperchain of two or more local DAGs. The following corollary generalizes the d-sepnode testing strategy established in Proposition 9.16 for the hyperstar. We denote the parents of x that the center agent shares with an adjacent agent A_k by $\pi_k(x)$.

Corollary 9.17 *Let x be a public node in a hyperstar local DAGs $\{G_i\}$, where all parents of x are public and each local DAG contains either x or some parents of x. The star has two or more terminals such that from the center to each terminal is a hyperchain of two or more local DAGs. Let the hyperstar be populated by agents with one for each local DAG.*

*After agents pass messages according to **CollectPublicParentInfoOnChain** (relative to x) from each terminal towards the center, the following hold:*

1. *If the center agent A_* receives 0 from any adjacent agent, x is a non-d-sepnode.*
2. *Otherwise, if 1 is received by A_* from each of any two adjacent agents, x is a non-d-sepnode.*
3. *Otherwise, if 1 is received by A_* from an adjacent agent A_j, and another adjacent agent A_k sharing the parents $\pi_k(x)$ of x with A_* exists such that $\pi_j(x) \not\supseteq \pi_k(x)$, then x is a non-d-sepnode.*
4. *Otherwise, x is a d-sepnode.*

Consider the hyperstar in Figure 9.25(a). It has the center at G_0 and three terminals G_1, G_2, and G_3. Suppose that it is populated by agents $A_* = A_0, A_1, A_2,$

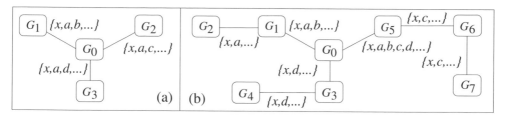

Figure 9.25: Parents $\pi(x)$ of a d-sepnode x shared by local DAGs in hyperstars.

and A_3. Each of A_1, A_2, and A_3 sends a message to A_0 according to **CollectPublicParentInfoOnChain**. Because each hyperchain has only two local DAGs, A_0 receives -1 from each other agent and concludes that x is a d-sepnode. Clearly, this is correct given that G_0 contains $\pi(x)$. The example shows that whenever a hyperstar has two local DAGs in each hyperchain from the center to a terminal, x is a d-sepnode.

As another example, consider another hyperstar in Figure 9.25(b). It has the center at G_0 and three terminals G_2, G_4, and G_7. After agents send messages according to **CollectPublicParentInfoOnChain**, A_0 receives -1 from each of A_1, A_3, and A_5 and concludes that x is a d-sepnode.

In Figure 9.26, the hyperstar has the same topology as that of Figure 9.25(b) but a different distribution of $\pi(x)$. Note how the parents of x are distributed along the hyperchain between G_0 and G_7. Agent A_0 receives -1 from A_1 and A_3 and receives 1 from A_5. According to Corollary 9.17, A_0 concludes that x is a d-sepnode. We see that both G_5 and G_6 contain $\pi(x)$.

Figure 9.27 shows a slightly different distribution of $\pi(x)$. Note the parents of x that are shared by G_3 and G_4. Agent A_0 receives -1 from A_1 and 1 from both A_3 and A_5. According to Corollary 9.17, A_0 concludes that x is not a d-sepnode. We see that no local DAG contains all parents of x.

In Figure 9.28, A_0 receives 1 from A_3, and -1 from A_1 and A_5. It concludes that x is not a d-sepnode by Corollary 9.17 because $\pi_1(x)$ and $\pi_3(x)$ are incomparable. We see that no local DAG contains both a and e.

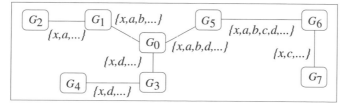

Figure 9.26: Parents $\pi(x)$ of a d-sepnode x shared by local DAGs in a hyperstar.

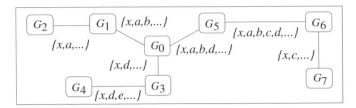

Figure 9.27: Parents $\pi(x)$ of a non-d-sepnode x shared by local DAGs in a hyperstar.

9.6.7 Cooperative d-sepnode Testing in a Hypertree

In Sections 9.6.4 and 9.6.6, we have considered d-sepnode testing where, for a given public variable x, each local DAG in a hyperchain or a hyperstar contains either x or some public parents of x. We now extend the results into the most general case. The extension consists of two aspects: First, we consider a general hypertree of local DAGs where some local DAGs contain neither x nor parents of x. Second, we allow local DAGs that do contain parents of x to form a (sub)hypertree, not just a hyperchain or hyperstar. We assume that **FindNonDsepnodeByPrivateParent** has been executed on the hypertree with **false** returned. By Proposition 9.7, no more than one local DAG contains private parents of x.

   We assume that the hypertree is populated by a set of agents. Immediately following we present recursive algorithms for each agent. An algorithm executed by an agent (denoted by A_0) is activated by a caller, which is either an adjacent agent (denoted by A_c) of A_0, or A_0 itself (see how Algorithm 9.13 is activated in Algorithm 9.14), or the system coordinator, which could be a human or an agent. If A_0 has adjacent agents other than A_c, denote them by $A_1, \ldots, A_k$. Their local DAGs are $G_c, G_0, G_1, \ldots, G_k$, respectively. We denote the parents of x shared by G_0 and G_i by $\pi_i(x)$ $(i = 1, \ldots, k)$ and those by G_0 and G_c by $\pi_c(x)$. Note that when A_0 is called by A_c, there may not be any other adjacent agent of A_0 because it is a terminal agent or no adjacent agent A_i $(i = 1, \ldots, k)$ that shares either x or some parents of x with A_0. For simplicity, we refer to all these cases as no adjacent agent A_i $(i = 1, \ldots, k)$ such that A_i shares either x or some parents of x. If an algorithm is activated by the system coordinator, we will use A_* to describe any agent thus selected.

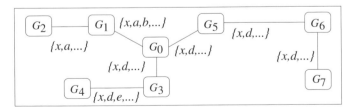

Figure 9.28: Parents $\pi(x)$ of a non-d-sepnode x shared by local DAGs in a hyperstar.

The first algorithm **CollectPublicParentInfo** collects information on the distribution of public parents of x by an inward propagation of messages among agents. The final returned message summarizes the distribution.

Algorithm 9.13 (CollectPublicParentInfo) *When the caller calls **CollectPublic-ParentInfo(x)** in A_0, it does the following:*

1. *Agent A_0 returns -1 if there is no adjacent agent A_i ($i = 1, \ldots, k$) such that A_i shares either x or some parents of x. Otherwise, it continues.*
2. *For each agent A_i ($i = 1, \ldots, k$) such that A_i shares either x or some parents of x, A_0 calls **CollectPublicParentInfo(x)** in A_i and collects a returned message m_i.*
3. *(a) If there exists $m_i = 0$ ($1 \leq i \leq k$), then A_0 returns 0.*
   (b) Otherwise, if there exist $m_i = 1$ and $m_j = 1$ ($1 \leq i, j \leq k$; $i \neq j$), then A_0 returns 0.
   (c) Otherwise, if there exists $m_i = 1$ ($1 \leq i \leq k$), then A_0 compares $\pi_i(x)$ with $\pi_j(x)$ for each j ($1 \leq j \leq k$ or $j = c$; $j \neq i$).
      If j is found such that $\pi_i(x) \not\supseteq \pi_j(x)$, A_0 returns 0. If not, A_0 returns 1.
   (d) Otherwise, continue.
4. *If caller is an adjacent agent A_c, A_0 compares $\pi_c(x)$ with $\pi_i(x)$ for each i ($1 \leq i \leq k$). If there exists i such that $\pi_c(x) \not\supseteq \pi_i(x)$, then A_0 returns 1. Otherwise, A_0 returns -1.*
5. *If caller is A_0 itself, A_0 returns -1.*

The following proposition says that if **CollectPublicParentInfo** is called on an agent whose local DAG contains all parent nodes of a d-sepnode, then its returned message must be -1.

Proposition 9.18 *Let a hypertree of local DAGs $\{G_i\}$ be populated by a set of agents. Let x be a public node and $\pi(x)$ be the set of all parents of x in the hypertree. Let A_* be the agent whose local DAG G_* contains $\pi(x)$.*
*If **CollectPublicParentInfo(x)** is called on A_*, A_* will return -1.*

Proof: Inasmuch as the local DAGs form a hypertree, those local DAGs, each of which contains either x or some parents of x, form a subtree. We only need to consider the processing by agents located in this subtree. Because **CollectPublic-ParentInfo(x)** is called on A_*, the subtree is viewed as being rooted at G_*. For each agent that is terminal in the subtree, it returns -1 (Case (1) of **CollectPublicParentInfo**). We show that for each nonterminal agent, if it receives only -1 as the result of calling **CollectPublicParentInfo** in its adjacent agents, then it will return -1 as well.

First, consider any nonterminal agent A_0 other than A_*. Because G_* contains $\pi(x)$, the public parent sequence of x from G_* to each terminal local DAG (with at least one other local DAG in between) is either identical or decreasing. Hence, when A_0 is called by A_c, it will find either $\pi_c(x) = \pi_i(x)$ or $\pi_c(x) \supset \pi_i(x)$. According to Case (4) of **CollectPublicParentInfo**, it will return -1. Therefore, each agent adjacent to A_* will return -1. As a consequence, by Case (5) of **CollectPublicParentInfo**, A_* will return -1.   □

The next proposition says that if **CollectPublicParentInfo** is called on an agent whose local DAG does not contain all parents of a public node, then it will not return -1.

Proposition 9.19 *Let a hypertree of local DAGs $\{G_i\}$ be populated by a set of agents. Let x be a public node and $\pi(x)$ be the set of all parents of x in the hypertree. Let $G_* = (V_*, E_*)$ be the only local DAG that may contain private parents of x and satisfies $V_* \cap \pi(x) \subset \pi(x)$.*

*Then when **CollectPublicParentInfo(x)** is called on A_*, A_* will not return -1.*

Proof: Because $V_* \cap \pi(x) \subset \pi(x)$, there exists $y \in \pi(x) \backslash V_*$ and a local DAG G_y that contains y and is the closest to G_* among such local DAGs exist. Because G_* is the only local DAG that may contain private parents of x, y is a public parent of x. Let G_i be the local DAG adjacent to G_y that contains y. See Figure 9.29 for an illustration, where $\pi_c(x)$ denotes the parents of x that A_0 shares with the adjacent agent A_c between A_y and A_* on the hypertree, and $\pi_i(x)$ represents the parents of x that A_y shares with the agent A_i. Note that it is possible that $G_* = G_c$.

Because G_c and G_y do not share y but G_i and G_y do, A_y will find either $\pi_c(x) \subset \pi_i(x)$ or $\pi_c(x) \bowtie \pi_i(x)$. According to **CollectPublicParentInfo**, -1 might be returned by A_y only under case 4. However, under case 4, given that $\pi_c(x) \not\supseteq \pi_i(x)$, A_y will return 1 to A_c. For each remaining agent on the chain from A_y to A_*, if it receives 1 or 0, it can only return 1 or 0 according to Cases (3a), (3b), or (3c). Hence, A_* will not return -1.   □

Combining Propositions 9.18 and 9.19, we have the following conclusion: If **CollectPublicParentInfo(x)** is called on A_* and we know that only G_* may contain private parents of x, then when -1 is returned, G_* must contain $\pi(x)$ and hence

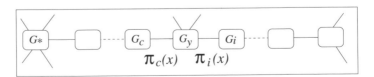

Figure 9.29: Illustration of proof for Proposition 9.19.

x is a d-sepnode. Otherwise, G_* cannot contain $\pi(x)$. This is summarized in Theorem 9.20, which follows directly from the two propositions.

Theorem 9.20 *Let a hypertree of local DAGs $\{G_i\}$ be populated by a set of agents. Let x be a public node and $\pi(x)$ be the set of all parents of x in the hypertree. Let $G_* = (V_*, E_*)$ be the only local DAG that may contain private parents of x. Let* **CollectPublicParentInfo(x)** *be called on A_*.*
Then $V_* \supset \pi(x)$ if and only if A_* returns -1.

If G_* does contain private parents of x, then Theorem 9.20 determines whether x is a d-sepnode with certainty. We will further explore this in Section 9.6.8. If all parents of x are public, when A_* returns -1, Theorem 9.20 says that it is also certain that x is a d-sepnode. However, if A_* does not return -1, it is possible that a local DAG other than G_* may contain $\pi(x)$. Hence, whether x is a d-sepnode is inconclusive in this case. Using the lessons learned from the analysis of hyperchains and hyperstars, if we know that there exists a hyperchain on the hypertree such that the public sequence of x along the hyperchain is of the wave type, we can be certain that no local DAG will contain $\pi(x)$ and x is not a d-sepnode. On the other hand, if no wave sequence of x exists, we can conclude that x is a d-sepnode. To explore this idea, we first consider intuitively how such a sequence might be present in a hypertree, as illustrated in Figure 9.30.

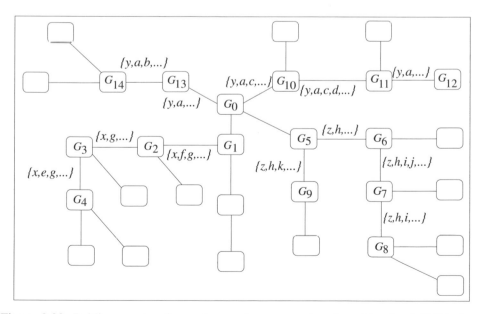

Figure 9.30: Public parents of non-d-sepnode x, y, and z shared by local DAGs in a hypertree.

Figure 9.30 shows a hypertree of local DAGs in which each DAG is represented as a box. For simplicity, only some local DAGs are labeled. The public parent sequences of only three variables (x, y, and z) are shown explicitly. The sequences for x, y, and z are of the wave type. Suppose that cooperative testing is activated at agent A_0 and hence that A_0 plays the role of root of the hypertree. The three parent sequences demonstrate three typical ways a parent sequence can be situated in the hypertree: The sequence for x is along a hyperchain from the root to a terminal:

$$\langle G_0, G_1, G_2, G_3, G_4, \ldots \rangle.$$

The sequence for y is split into two hyperchains each from the root to a terminal:

$$\langle G_0, G_{13}, G_{14}, \ldots \rangle \quad \text{and} \quad \langle G_0, G_{10}, G_{11}, G_{12} \rangle.$$

The sequence for z is split into two hyperchains each from a nonroot to a terminal:

$$\langle G_5, G_6, G_7, G_8, \ldots \rangle \quad \text{and} \quad \langle G_5, G_9, \ldots \rangle.$$

The following proposition establishes that when a wave parent sequence of x is present, if **CollectPublicParentInfo(x)** is called on A_*, it will return 0.

Proposition 9.21 *Let a hypertree of local DAGs $\{G_i\}$ be populated by a set of agents. Let x be a public node without private parents such that there exists a hyperchain along which its public parent sequence is of the wave type. Let G_* be a local DAG that contains either x or some parents of x and **CollectPublicParentInfo(x)** be called on A_*.*
Then A_* will return 0.

Proof: We show that if a wave public parent sequence of x occurs in any one of the three possible ways (denoted by T_1, T_2, and T_3) demonstrated in Figure 9.30, A_* will return 0.

[T_1] Consider a wave public parent sequence s of x that occurs along a hyperchain from A_* to a terminal agent A_t (e.g., the parent sequence for x in Figure 9.30). Since s is a wave sequence, from A_t to A_* along the hyperchain, there exist G_i first and then G_j (see Figure 9.31) such that $\pi_i^-(x) \subset \pi_i^+(x)$ or $\pi_i^-(x) \bowtie \pi_i^+(x)$ and $\pi_j^-(x) \supset \pi_j^+(x)$ or $\pi_j^-(x) \bowtie \pi_j^+(x)$, where $\pi_i^-(x)$ denotes the parents of x that G_i

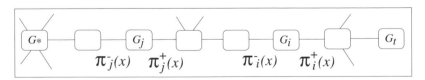

Figure 9.31: Illustration for proof of Proposition 9.21.

shares with the adjacent local DAG closer to A_* and $\pi_i^+(x)$ with that farther from A_*, and $\pi_j^-(x)$ and $\pi_j^+(x)$ are those relative to G_j. We claim that either 1 or 0 will be returned by A_i when **CollectPublicParentInfo(x)** is called on A_i, and 0 will be returned by A_j.

Agent A_i will return 1 or 0 in the Cases of (3a), (3b), and (3c) of **CollectPublicParentInfo**. Agent A_i will return 1 in Case (4) because $\pi_i^-(x) \not\supseteq \pi_i^+(x)$. Given that A_i returns 1 or 0, the agents between A_i and A_j can only return 1 or 0, for only Cases (3a), (3b), and (3c) are applicable. In the Cases of (3a) and (3b), A_j will return 0. In the Case of (3c), where A_j receives 1 from its adjacent agent on s owing to $\pi_j^-(x) \not\subseteq \pi_j^+(x)$, A_j will return 0.

$[T_2]$ Consider a wave parent sequence s that runs along two joined hyperchains c_1 and c_2. The hyperchain c_1 runs from an agent A_k to a terminal agent and c_2 runs from A_k to another terminal agent, where A_k is the agent closest to A_* among all agents on the two hyperchains (e.g., the parent sequence for z in Figure 9.30, where $A_k = A_5$). We assume that s is a wave sequence, but c_1 and c_2 are not, for if any one of them were, it would be detected, as in case T_1. Hence, it suffices to consider the following two cases:

1. There exist an agent A_i exists on c_1 and another agent A_j on c_2 that satisfy the following: For A_i, either $\pi_i^-(x) \subset \pi_i^+(x)$ or $\pi_i^-(x) \bowtie \pi_i^+(x)$ holds, where $\pi_i^-(x)$ denotes the parents of x that G_i shares with the adjacent local DAG closer to A_k and $\pi_i^+(x)$ with that farther from A_k. For A_j, either $\pi_j^-(x) \subset \pi_j^+(x)$ or $\pi_j^-(x) \bowtie \pi_j^+(x)$ holds. See Figure 9.32 for an illustration.
2. Agent A_i is identical to case 1 above, but $G_j = G_k$ such that either $\pi_1(x) \subset \pi_2(x)$ or $\pi_1(x) \bowtie \pi_2(x)$ holds, where $\pi_1(x)$ represents the parents of x that G_k shares with the adjacent local DAG on c_1 and $\pi_2(x)$ with that on c_2. See Figure 9.33 for an illustration.

For case 1, when **CollectPublicParentInfo** is called on A_i, it will return 1 or 0 as argued in T_1, and so will A_j. Hence, agents between A_i and A_k will return 1 or 0,

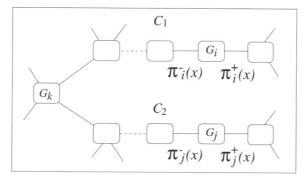

Figure 9.32: Hyperchains c_1 and c_2.

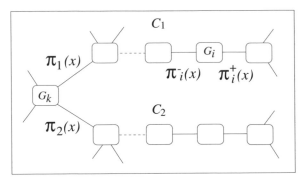

Figure 9.33: Hyperchains c_1 and c_2, where $G_j = G_k$.

for only (3a), (3b), and (3c) of **CollectPublicParentInfo** are applicable. The same is true for agents between A_j and A_k. As a result, A_k will receive 1 or 0 from its adjacent agents in c_1 and c_2. According to Case (3a) and Case (3b), A_k will return 0.

For case 2, A_k will receive 1 or 0 from its adjacent agents in c_1. It will return 0 if it receives 0 by Case (3a). If it receives 1 from its adjacent agent on c_1 and some other adjacent agents, it will return 0 by Case (3b). If it receives 1 only from its adjacent agent on c_1, then by Case (3c) it will return 0 because $\pi_1(x) \not\supseteq \pi_2(x)$.

[T_3] Consider a wave parent sequence s that runs along two hyperchains c_1 and c_2 joined at A_*. When **CollectPublicParentInfo(x)** is called on A_*, it will return 0 for reasons similar to argument presented in T_2.   □

When a wave parent sequence of x exists, no local DAG may contain $\pi(x)$; hence, x is not a d-sepnode. Therefore, Proposition 9.21 establishes that whenever x is not a d-sepnode, **CollectPublicParentInfo(x)** called on A_* will return 0. The proposition, however, does not tell us when x is a d-sepnode whether it is still possible that 0 will be returned. The following proposition answers this question. It asserts that in such a case, 0 will never be returned. Thus, a returned 0 uniquely identifies x as a non-d-sepnode.

Proposition 9.22 *Let a hypertree of local DAGs $\{G_i\}$ be populated by a set of agents. Let x be a d-sepnode without private parents. Let G_* be a local DAG that contains either x or some parents of x and **CollectPublicParentInfo(x)** be called on A_*.*

Then A_* will not return 0.

Proof: Because x is a d-sepnode, there exists a local DAG G_j that contains $\pi(x)$. We have either $G_j = G_*$ or $G_j \neq G_*$. When $G_j = G_*$, by Proposition 9.18, A_* will return -1.

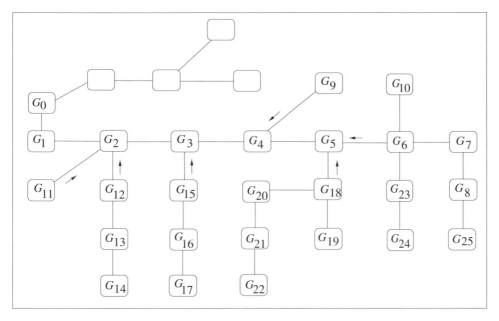

Figure 9.34: Proof of Proposition 9.22.

In the case $G_j \neq G_*$, we analyze the processing performed by agents located on the hyperchain between G_* and G_j inclusive. In Figure 9.34, if we assume $G_* = G_2$ and $G_j = G_5$, these agents include A_2 through A_5. For any one of them, say A_k, we consider the return message when **CollectPublicParentInfo(x)** is called by A_k in one of its adjacent agents that is not on this hyperchain. Because each parent sequence from G_k to a downstream terminal (relative to G_*) is identical or decreasing, the message -1 will be returned to A_k on the basis of Cases (1) and (4) of **CollectPublicParentInfo(x)**. For example, A_5 receives -1 from A_6 and A_{18}, A_3 receives -1 from A_{15}, and A_2 receives -1 from A_{11} and A_{12}. Hence, each agent on the hyperchain from G_j to G_* will receive -1 when it calls **CollectPublicParentInfo(x)** in an agent not on the hyperchain.

The parent sequence from G_* to G_j is identical or increasing because G_j contains $\pi(x)$. When **CollectPublicParentInfo(x)** is called on A_j, it receives -1 only, as argued above. It will return 1 or -1 on the basis of Case (4). The caller agent of A_j receives -1 from each adjacent agent with the possibility of receiving 1 from A_j only. If only -1 is received, it returns 1 or -1 as A_j. If a unique 1 is received, it returns 1 based on Case (3c). Applying this analysis recursively to the other agents on the hyperchain from G_j to G_*, we conclude that each of them returns 1 or -1. Hence, A_* will not return 0. □

With the understanding of the relation between the return message of **Collect-PublicParentInfo(x)** and the parent distribution of x, we present the following recursive algorithm **FindNonDsepnodeByPublicParent** to verify if a hypertree satisfies the d-sepset condition under the assumption that for each public node all its parents are also public.

Algorithm 9.14 (FindNonDsepnodeByPublicParent) *When FindNonDsepnodeByPublicParent is called on an agent A_0, it does the following:*

1. *Agent A_0 returns **false** if it has no adjacent agent except A_c. Otherwise, it continues.*
2. *For each public node x in G_0 such that x is not shared by A_c, A_0 calls on itself **CollectPublicParentInfo(x)**. If 0 is returned, A_0 returns **true**. Otherwise, it continues.*
3. *For each agent A_i ($i = 1, \ldots, k$), A_0 calls **FindNonDsepnodeByPublicParent** in A_i. If A_i returns **true**, then A_0 returns **true**.*
4. *Agent A_0 returns **false**.*

Note the restriction "x is not shared by A_c" in the second step above. It ensures that **CollectPublicParentInfo(x)** is executed exactly once and is activated when the distributed processing reaches the first agent on the hypertree that contains x.

The following theorem shows that if **FindNonDsepnodeByPublicParent** is called on an agent A_*, the return flag will reflect correctly whether each shared node with only public parents is a d-sepnode.

Theorem 9.23 *Let a hypertree of local DAGs $\{G_i\}$ be populated by a set of agents. For each node x shared by two or more local DAGs, all parents of x are public. Let an agent A_* be arbitrarily chosen and **FindNonDsepnodeByPublicParent** be called on A_*.*

*If A_* returns **true**, then at least one shared node is a non-d-sepnode. Otherwise, all shared nodes are d-sepnodes.*

Proof: We show that if there exists a non-d-sepnode, A_* will return **true**, and if all public nodes are d-sepnodes, A_* will return **false**. For each public node x with public parents, those local DAGs, each containing either x or some public parents of x, form a subhypertree. This is due to the hypertree property (Definition 6.8). Because the processing of **FindNonDsepnodeByPublicParent** propagates through the hypertree outwards from A_*, **CollectPublicParentInfo(x)** will be called on an agent A_0 such that G_0 is located in the subhypertree and is the closest to G_* among those on the subhypertree and containing x.

Note that G_0 may not be the closest local DAG to G_* on the subhypertree because another G_y may contain a parent y of x, not contain x, and be between G_0 and G_* on the hypertree. It is also possible that still another G'_y may contain y, may not contain x, and the hyperchain from G_* to G'_y does not even include G_0. The effect of these local DAGs is to be considered in Exercise 9. In the following, we assume that G_0 is the closest local DAG to G_* on the subhypertree.

If there exists a non-d-sepnode x, there must be a hyperchain on the hypertree along which the public parent sequence of x is a wave formation. By Proposition 9.21, when **CollectPublicParentInfo(x)** is called on A_0, 0 will be returned. This will cause A_0 to return **true** in **FindNonDsepnodeByPublicParent** and eventually cause A_* to return **true**.

If each public node is a d-sepnode, then when **CollectPublicParentInfo(x)** is called on an agent A_0 it will not return 0 according to Proposition 9.22. Because none of the calls of **CollectPublicParentInfo** returns 0, A_* will return **false** as the result of **FindNonDsepnodeByPublicParent**. □

Theorem 9.23 shows that **FindNonDsepnodeByPublicParent** solves the d-sepnode verification problem if all shared nodes have only public parents. In the following section, we present algorithms for d-sepnode verification when some public nodes have private parents.

9.6.8 Cooperative Verification of d-sepset Condition

In this section, we first extend **FindNonDsepnodeByPublicParent** to d-sepnode verification of public nodes with private parents. We then assemble the algorithms developed so far for verification of the d-sepset condition in a general hypertree DAG union.

Recall from Section 9.6.3 that, if a d-sepnode has private parents, all such parents must be contained in a single local DAG. According to Theorem 9.20, if a public node x has private parents in a local DAG G_*, then whether x is a d-sepnode can be determined by calling **CollectPublicParentInfo(x)** in A_* and checking the return message.

Suppose that a single local DAG contains private parents of x. When the following **FindNonDsepnodeByHub** is called on an agent whose local DAG contains either x or some parents of x, a recursive search is performed through the hypertree for the single local DAG that contains the private parents of x. When the local DAG is found, **CollectPublicParentInfo(x)** is called on the corresponding agent to verify if x is a d-sepnode.

Algorithm 9.15 (FindNonDsepnodeByHub) *When* **FindNonDsepnodeBy-Hub(x)** *is called on an agent A_0 that contains either x or some parents of x, it does the following:*

1. If G_0 contains no private parent of x, then A_0 checks if there exists A_i $(i = 1, \ldots, k)$ that contains either x or some public parents of x.

*If found, then for each such A_i, A_0 calls **FindNonDsepnodeByHub(x)** in A_i. If any such A_i returns **true**, then A_0 returns **true**. If each such A_i returns **false**, A_0 returns **false**. If no such A_i is found, then A_0 returns **false**.*

2. *If G_0 contains private parents of x, then A_0 calls **CollectPublicParentInfo(x)** on itself. If -1 is returned, A_0 returns **false**. Otherwise, A_0 returns **true**.*

The processing activated by **FindNonDsepnodeByHub(x)** may terminate in two ways. If A_0 contains no private parent of x and its adjacent agents (except A_c) do not contain either x or some public parents of x, then **FindNonDsepnodeByHub(x)** will terminate. When an agent A_0 containing private parents of x is reached, **CollectPublicParentInfo(x)** will be recursively activated. **CollectPublicParentInfo(x)** will terminate when an agent A_i is activated, and A_i and its adjacent agents (except its caller agent) do not contain either x or some public parents of x.

The following **FindNonDsepnode** checks if each public node x is a d-sepnode by propagating processing outwards through the hypertree. It does this in the same way as **FindNonDsepnodeByPublicParent** if x has no private parents. Otherwise, the checking is performed through **FindNonDsepnodeByHub**. In the algorithm, a record $count(x)$ is used. Recall that this record is saved by **FindNonDsepnodeByPrivateParent** in Section 9.6.2. If $count(x) = 0$, x has no private parents, and if $count(x) = 1$, x has private parents in exactly one local DAG.

Algorithm 9.16 (FindNonDsepnode) *When **FindNonDsepnode** is called on an agent A_0, it does the following:*

1. *Agent A_0 returns **false** if it has no adjacent agent except A_c. Otherwise, it continues.*
2. *For each public node x in G_0 such that x is not shared by A_c, A_0 checks $count(x)$.*
    *If $count(x) = 0$, A_0 calls on itself **CollectPublicParentInfo(x)**. If 0 is returned, A_0 returns **true**. Otherwise, it continues.*
    *If $count(x) = 1$, A_0 calls on itself **FindNonDsepnodeByHub(x)**. If **true** is returned, A_0 returns **true**. Otherwise, it continues.*
3. *For each agent A_i ($i = 1, \ldots, k$), A_0 calls **FindNonDsepnode** on A_i. If A_i returns **true**, then A_0 returns **true**.*
4. *A_0 returns **false**.*

The following **VerifyDsepset** is executed by the system coordinator to verify that the entire hypertree DAG union satisfies the d-sepset condition. After the root agent A_* is selected, **FindNonDsepnodeByPrivateParent** is used to determine if there is any public node for which more than one local DAGs contain its private parents. If the result is positive (**false** returned), a side effect of the algorithm is that for each public node x a record $count(x)$ is saved in an agent that is the closest to A_*

on the hypertree among agents that contain x. This record is used by the following operation **VerifyDsepset**, which completes the d-sepset verification.

Algorithm 9.17 (VerifyDsepset) *Let a hypertree DAG union G be populated by multiple agents with one at each hypernode. The system coordinator does the following:*

1. *Choose an agent A_* arbitrarily.*
2. *Call **FindNonDsepnodeByPrivateParent** in A_*. If **true** is returned, then conclude that G violates the d-sepset condition. Otherwise, continue.*
3. *Call **FindNonDsepnode** in A_*. If **true** is returned, then conclude that G violates the d-sepset condition. Otherwise, conclude that G satisfies the d-sepset condition.*

The following theorem establishes that **VerifyDsepset** verifies the d-sepset condition correctly.

Theorem 9.24 *Let a hypertree DAG union G be populated by multiple agents. After **VerifyDsepset** is executed in G, it concludes correctly with respect to whether G satisfies the d-sepset condition.*

Proof: If G does not satisfy the d-sepset condition, then there exists a public node x that is not a d-sepnode. It may disqualify as a d-sepnode in three possible ways: It is possible that two local DAGs contain private parents of x, which will be detected by **FindNonDsepnodeByPrivateParent** according to Proposition 9.7. It is also possible that a single local DAG G_0 contains private parents of x but does not contain $\pi(x)$. In this case, **FindNonDsepnode** will find $count(x) = 1$ and call **FindNonDsepnodeByHub(x)**. **FindNonDsepnodeByHub(x)** will locate G_0 and call **CollectPublicParentInfo(x)** in A_0. The non-d-sepnode will be detected according to Theorem 9.20. Finally, it is possible that x has only public parents but that they form a wave sequence. In this case, **FindNonDsepnode** will find $count(x) = 0$ and call **CollectPublicParentInfo(x)**. The non-d-sepnode will be detected according to Theorem 9.23.

If none of the above occurs, then all public nodes are d-sepnodes by Theorem 9.20 and Theorem 9.23. **FindNonDsepnode** will return **false**, which causes **VerifyDsepset** to conclude accordingly. $\qquad\square$

We illustrate the execution of **VerifyDsepset** with the hypertree DAG union in Figure 9.35. It is a modification of the hypertree DAG union in Figure 9.19 with the arc (q, j) in G_4 reversed.

1. Suppose A_4 is selected by the system coordinator as A_*. **FindNonDsepnodeByPrivateParent** is then called on A_4.

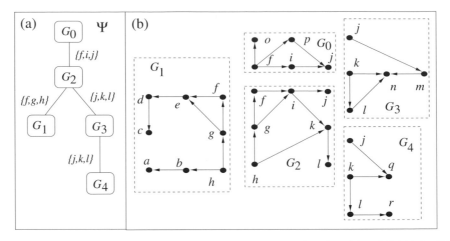

Figure 9.35: A hypertree DAG union with the hypertree in (a) and local DAGs in (b).

2. Agent A_4 calls on itself **CollectPrivateParentInfo(j)**. **CollectPrivateParentInfo(j)** is called subsequently in A_3, A_2, and A_0. Eventually **counter$_3$** $= 1$ is returned to A_4. It saves $count(j) = 1$.

   Agent A_4 calls on itself **CollectPrivateParentInfo(k)**. The call is propagated to A_3 and then A_2, and A_4 eventually saves $count(k) = 0$.

   CollectPrivateParentInfo(l) is then called by A_4 on itself and propagated to A_3 and A_2. As a result, $count(l) = 0$ is saved at A_4.

3. Agent A_4 calls **FindNonDsepnodeByPrivateParent** in A_3. Because each of the public nodes in A_3 is shared with A_4, A_3 does not activate **CollectPrivateParentInfo**. Instead, it calls **FindNonDsepnodeByPrivateParent** in A_2.

4. Agent A_2 calls on itself **CollectPrivateParentInfo(f)**. **CollectPrivateParentInfo(j)** is called subsequently on A_0 and A_1. As a result, $count(f) = 0$ is saved at A_2.

   Agent A_2 calls **CollectPrivateParentInfo(i)** on itself and then in A_0. The result $count(i) = 0$ is saved at A_2.

   Agent A_2 calls on itself **CollectPrivateParentInfo(g)** and then calls on A_1 with $count(g) = 0$ saved.

   Then A_2 calls on itself **CollectPrivateParentInfo(h)** and subsequently in A_1 with $count(h) = 0$ saved.

5. Agent A_2 calls **FindNonDsepnodeByPrivateParent** on A_0 and A_1, which return **false** immediately. Then A_2 returns **false** to A_3, which in turn returns to A_4.

6. Because A_4 returns **false** as the result of **FindNonDsepnodeByPrivateParent** called on it, the verification continues with the call **FindNonDsepnode** in A_4.

7. Agent A_4 checks $count(j) = 1$ and calls **FindNonDsepnodeByHub(j)** on itself. Since A_4 has no private parents of j, it calls **FindNonDsepnodeByHub(j)** in A_3. The call propagates to A_2 and then to A_0.

8. Agent A_0 has a private parent p of j and hence calls **CollectPublicParentInfo(j)** on itself. **CollectPublicParentInfo(j)** is then activated in A_2, A_3, and A_4.

Agent A_4 returns -1 to A_3. Because A_3 shares no parent of j with A_2 and A_4, A_3 returns -1 to A_2. Given that A_2 shares the parent i of j with A_0, A_2 returns -1 to A_0, which terminates **CollectPublicParentInfo(j)** with -1 returned.

9. Agent A_0 then returns **false** to A_2 as the result of the call **FindNonDsepnodeByHub(j)** from A_2. Subsequently A_2 returns **false** to A_3, which in turn returns **false** to A_4, which terminates **FindNonDsepnodeByHub(j)** with **false** returned.

10. Continuing the processing in **FindNonDsepnode**, A_4 next checks $count(k) = 0$. It calls **CollectPublicParentInfo(j)** on itself. **CollectPublicParentInfo(j)** is then activated in A_3, A_2, A_1, and A_0.

11. Agents A_0 and A_1 return -1 to A_2, which returns 1 to A_3 because its interface with A_0 contains the parent i of k, its interface with A_1 contains the parent h of k, and its interface with A_3 contains no parent of k. Agent A_3 passes 1 to A_4, which returns 1 as the result of calling **CollectPublicParentInfo(j)** on itself.

12. Continuing the processing in **FindNonDsepnode**, A_4 next checks $count(l) = 0$. It calls **CollectPublicParentInfo(l)** on itself. **CollectPublicParentInfo(l)** is then activated in A_3 and A_2. Eventually, -1 is returned to A_4.

13. Agent A_4 calls **FindNonDsepnode** in A_3. Because each variable that A_3 shares with A_2 is also shared with A_3, A_3 calls **FindNonDsepnode** in A_2.

14. In response, A_2 checks $count(f) = 0$ and calls **CollectPublicParentInfo(f)** on itself. Agent A_2 then calls **CollectPublicParentInfo(f)** on A_0 and A_1. As a result, A_2 eventually returns -1.

15. Continuing the processing in **FindNonDsepnode**, A_2 checks $count(i) = 0$ and calls **CollectPublicParentInfo(i)** on itself. The call is propagated to all other agents. Agents A_0 and A_1 return -1 to A_2. Then A_4 returns -1 to A_3, which returns -1 to A_2. Hence, A_2 terminates **CollectPublicParentInfo(i)** with -1 returned.

16. Continuing the processing in **FindNonDsepnode**, A_2 checks $count(g) = 0$ and calls **CollectPublicParentInfo(g)** on itself. The call is activated in A_1 because A_2 and A_1 share g and its parent h. Eventually, A_2 returns -1 from **CollectPublicParentInfo(g)**.

17. Agent A_2 checks $count(h) = 0$ and calls **CollectPublicParentInfo(h)** on itself. The call is activated in A_1, and eventually A_2 returns -1 from **CollectPublicParentInfo(h)**.

18. Agent A_2 calls **FindNonDsepnode** in A_0 and A_1, which return **false** immediately. Then A_2 returns **false** as the result of a call of **FindNonDsepnode** from A_3, which in turn returns **false** to A_4.

19. Finally, A_4 returns **false** from the call of **FindNonDsepnode** on it by the system coordinator. It is concluded that the DAG union satisfies the d-sepset condition.

9.7 Complexity of Cooperative d-sepset Testing

We show that multiagent cooperative verification by **VerifyDsepset** is efficient. We use the following notations:

- t: the maximum cardinality of a node adjacency in a local DAG.
- k: the maximum number of nodes in an agent interface.

- s: the maximum number of agents adjacent to any given agent on the hypertree.
- n: the total number of agents in the hypertree DAG union.

First, we consider **FindNonDsepnodeByPrivateParent**. Its computation is dominated by calling **CollectPrivateParentInfo.** Each agent may call **CollectPrivateParentInfo** $O(k\ s)$ times – one for each shared node. Each call may propagate to $O(n)$ agents. To examine whether a shared node has private parents in a local DAG takes $O(t)$ time. Hence, the total time complexity of **FindNonDsepnodeByPrivateParent** is $O(n^2\ k\ s\ t)$.

Next, we consider **FindNonDsepnode**. Its computation time is dominated by the call of **CollectPublicParentInfo** either directly or through **FindNonDsepnodeByHub**. Each agent may call **CollectPublicParentInfo** $O(k\ s)$ times. Each call may propagate to $O(n)$ agents. When processing public parent sequence information, an agent may compare $O(s)$ pairs of agent interfaces. Each comparison examines $O(k^2)$ pairs of shared nodes. Hence, the total time complexity of **FindNonDsepnode** is $O(n^2\ k^3\ s^2)$. The overall complexity of **VerifyDsepset** is then

$$O(n^2\ (k^3\ s^2 + k\ s\ t)),$$

and the computation is efficient.

9.8 Bibliographical Notes

The issue of subdomain verification was inspired during a discussion with Finn Jensen while I was visiting Aalborg University in 1998. The solution was presented in Xiang, Olesen, and Jensen [93]. Multiagent cooperative verification of DAG union acyclicity was proposed by Xiang [87]. Xiaoyun Chen assisted in refining multiagent d-sepset verification.

9.9 Exercises

1. Determine from W_0 through W_3 in Section 9.1 whether a hypertree can be constructed from V_0 through V_3.
2. Modify the algorithm **CollectFamilyInfo** to make it more efficient using the idea hinted in Section 9.5.4.
3. Trace the execution of **FindNonDsepnodeByPrivateParent** for the hypertree DAG union in Figure 9.19 with the arc (q, j) reversed.
4. Determine the type of public parent sequence of x on the hyperchain of local DAGs in Figure 9.24.
5. Prove Corollary 9.13.
6. Prove Proposition 9.14.

7. Let x be a public node in a hyperstar DAG union, where all parents of x are public and each local DAG contains either x or some parents of x. There are exactly two local DAGs from the center of the star to each terminal inclusive. Prove that x is a d-sepnode.

8. Prove Theorem 9.20.

9. In the second step of algorithm **FindNonDsepnodeByPublicParent**, A_0 calls on itself **CollectPublicParentInfo(x)** if x is not shared by A_c. It is possible that A_c contains some parents of x. Hence, **CollectPublicParentInfo(x)** will not include these parents of x in its examination of the public parent sequences of x.

   Analyze the consequence of this omission.

10

Looking into the Future

In Chapters 6 through 9, we studied in detail why a set of agents over a large and complex domain should be organized into an MSBN and how. We studied how they can perform probabilistic reasoning exactly, effectively, and distributively. In this chapter, we discuss other important issues that have not yet been addressed but will merit research effort in the near future.

10.1 Multiagent Reasoning in Dynamic Domains

Practical problem domains can be *static* or *dynamic*. In a static domain, each domain variable takes a value from its space and will not change its value with time. Hence, at what instant in time an agent observes the variable makes no difference. On the other hand, in a dynamic domain, a variable may take different values from its space at different times. The temperature of a house changes after heating is turned on. The pressure of a sealed boiler at a chemical plant increases after the liquid inside boils. A patient suffers from a disease and recovers after the proper treatment. A device in a piece of equipment behaves normally until it wears out. Dynamic domains are more general, and a static domain can be viewed as a snapshot of a dynamic domain at a particular instant in time or within a time period in which the changes of variable values are ignorable.

A Bayesian network can be used to model static and dynamic domains. In general, given a dynamic domain, a domain variable $v(t)$ whose value changes with time t can be modeled by a sequence of variables $v(0), v(1), v(2), \ldots, v(i), \ldots,$ where i indexes a time instant. The collection of variables corresponding to time i and their dependence relations can be modeled by a Bayesian (sub)network. The variable $v(i)$ may also be dependent on some variable $x(j)$, where $j < i$. Such dependence is called *temporal dependence* and can be represented by an arc from $x(j)$ to $v(i)$. In this way, all Bayesian subnetworks corresponding to different time

274

instants are connected into one Bayesian network. Probabilistic reasoning can then be performed as usual.

However, if it is necessary to model and reason about a domain over an unbounded period of time, the preceding method is impractical, for the Bayesian network will grow to an unbounded size. To handle this, the first common step is to assume that the dynamic domain is *stationary*. That is, (1) the dependence relations among variables at time i are identical to those at j, and (2) the temporal dependence relations between variables at times i and j are identical to those between variables at times $i + k$ and $j + k$, where k is a positive integer. Condition (1) implies that all Bayesian subnets, one for each time instant, are isomorphic. Condition (2) means that if there is an arc from $x(j)$ to $v(i)$, then there is an arc from $x(j + k)$ to $v(i + k)$. The second common step is to assume that the dynamic domain is *Markovian*. That is, the state of the domain at time $i + 1$ is independent of the states at $0, 1, \ldots, i - 1$ given the state at i. This assumption implies that the Bayesian subnet at time i has only incoming arcs from the Bayesian subnet at time $i - 1$ and has only outgoing arcs to the Bayesian subnet at time $i + 1$.

Given a dynamic domain that is both stationary and Markovian, the modeling task and reasoning task are much simplified. It is now only necessary to model the domain at a particular time i and its dependence on time $i - 1$. The sequence of identical Bayesian subnets connected by temporal arcs is called a *dynamic Bayesian network* (DBN). To reason about the domain using a DBN, an agent starts with the Bayesian subnet at time 0. When an observation is available, the agent updates belief as usual and propagates its belief to the Bayesian subnet at time 1. Because the Bayesian subnet at time 1 now contains all the relevant historic information at time 0, the Bayesian subnet at time 0 need not be maintained. Repeating this operation, the agent can reason about the domain over an unbounded time period by maintaining no more than two Bayesian subnets at consecutive time instants.

We would like multiple agents to be able to reason about a large dynamic domain over an unbounded time period in a similar fashion. The multiagent MSBN framework presented in Chapters 6 through 9 allows modeling and reasoning about a dynamic domain over a finite time $(i, i + 1, \ldots, i + k)$ between instants i and $i + k$, where $k \geq 0$. Figure 10.1(a) shows the structure of an MSBN over one time instant ($k = 0$), and (b) shows the structure of an MSBN over two time instants ($k = 1$). Note the temporal dependence signified by the arcs (a_0, a_1) and (g_0, g_1).

One way to apply the framework to an unbounded time period directly is to let agents initialize to prior belief at the beginning of each k-instant period. The limitation of this method is that the relevant information acquired during the previous k-instant period cannot be used.

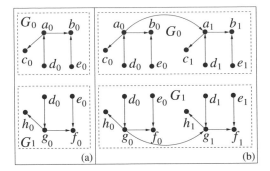

Figure 10.1: MSBNs of a trivial dynamic domain. (a) Over one time instant. (b) Over two time instants.

Ideally, we would like multiple agents to be able to benefit from each other's knowledge up to the relevant history. Unfortunately, exact multiagent probabilistic reasoning over an unbounded time period is impractical. Clearly, it is impractical to use the preceding representation and increase the value k unboundedly. We consider another alternative following the idea of dynamic Bayesian networks. Let us assume that the dynamic domain is stationary and Markovian. We would like each agent to maintain its belief on its subdomain for the current time instant. For example, at time $t = 0$, we would like agents A_0 and A_1 to maintain the subnets G_0^0 and G_1^0 in Figure 10.2(a), respectively. At $t - 1$, we would like them to maintain the subnets G_0^1 and G_1^1 in (b). By Proposition 6.4, each agent interface should render subdomains in the two induced (hyper)subtrees conditionally independent. Otherwise, exact probabilistic reasoning through message passing is impossible in general. In this example, the issue becomes whether

$$I(\{a_1, b_1, c_1\}, \{d_1, e_1\}, \{f_1, g_1, h_1\})$$

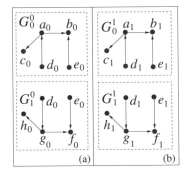

Figure 10.2: (a) Subnets to be maintained by agents at time $t = 0$. (b) Subnets to be maintained by agents at time $t = 1$.

holds. From Figure 10.1(b), we can see that, although

$$I(\{a_1, b_1, c_1\}, \{d_0, e_0, d_1, e_1\}, \{f_1, g_1, h_1\})$$

holds,

$$I(\{a_1, b_1, c_1\}, \{d_1, e_1\}, \{f_1, g_1, h_1\})$$

does not because the path

$$\rho = \langle a_1, a_0, d_0, g_0, g_1 \rangle$$

is rendered open by $\{d_1, e_1\}$ according to d-separation. Therefore, message passing between A_0 and A_1 by exchanging belief over their interface $\{d_1, e_1\}$ does not ensure correct belief updating in each agent. This demonstrates that exact multiagent probabilistic reasoning over unbounded time periods cannot be achieved by maintaining agent belief over a finite time.

In summary, exact multiagent probabilistic reasoning over unbounded time periods is impractical. Therefore, approximations using heuristic methods are necessary. Many alternatives are yet to be explored.

10.2 Multiagent Decision Making

Throughout the book, we have concentrated on how multiple agents can organize themselves in reasoning about the state of a large and complex domain. If an agent believes that a device in a piece of equipment is faulty, the agent should then either sound an alarm or replace the device with a backup. If a heating agent in a smart house believes that the people in the house have gone to work or school and only pets are wandering in the house, the agent can reduce the heating to save energy. As we understand better how agents can cooperate to reason about their uncertain environment, a natural step forward is to investigate how agents can make decisions and take actions in the environment.

According to the decision theory, an agent should choose among alternative actions based on its belief about the state of the domain, the possible consequences of actions, and its preference about the consequences. The preference can be represented by a value distribution over the states.

In a multiagent system, an agent A_i has at least three types of actions to take. To update its belief, A_i may choose to observe a local variable, which may incur a cost. We refer to this action type as *local observation*. Alternatively, A_i may choose to communicate with other agents so that it can benefit from others' observations. We refer to this action type as *communication*. Local observation and communication

are information-gathering actions, and they do not change the state of the domain.[1]
Finally, A_i may take actions to change the domain state such as replacing a device
in a piece of equipment with a backup or reducing the heating in a smart house. We
refer to this action type as *state transition*.

When deciding on state transition actions, A_i may deliberate its actions stepwise.
That is, it decides on a state transition action, takes it, performs a local observation
or communication to update its belief about the new state of the domain, and decides
on the next state transition action. Alternatively, A_i may plan a sequence of state
transition actions before any one is executed. The sequence may interleave state
transition actions with local observations and communications.

The research issues include the following:

- *Influence diagrams* (Howard and Matheson [26], Oliver and Smith [45]), *Markov decision*
  processes (MDP) (Puterman [57]), and *partially observable Markov decision processes*
  (POMDPs) (Smallwood and Sondik [66], Monahan [41]) extend the Bayesian networks
  and Markov chains to allow representation of alternative actions and user preferences.
  Can similar extensions be made to MSBNs so that stepwise decision making can be
  supported? Can planning a sequence of actions be similarly supported?

  It has been shown (Bernstein, Zilberstein, and Immerman [3]) that the computation for
  solving distributed POMDPs has a very high complexity (probably more than exponential). Multiagent probabilistic inference using an MSBN representation is also based on
  partial observations. It is distributed, exact, and efficient (when the DAG union is sparse).
  Therefore, extensions of the MSBN framework provide good candidates for tackling
  problems intended by POMDPs.

- How should the preference of cooperative agents be represented distributively?

- Although cooperative agents work for a single principal, each is limited by its resources.
  Conflicts may occur when an agent needs to perform multiple activities while it can only
  perform one at a time. For example, A_i may be asked to perform a communication by
  another agent while A_i is trying to take a state transition action. How should such conflicts
  be resolved in a decision-theoretic fashion?

- In deciding local observation and communications actions, an agent needs to trade the
  expected value of an information-gathering action with its cost. Trade-off between the
  information value of observation and its cost has been studied under the single-agent
  paradigm. The possibility of trading local observation with communication opens new
  opportunities and new issues.

- After a state transition action is performed, the beliefs of agents based on past observations
  may no longer be valid. For instance, after a faulty device is replaced, observations on
  variables that depend on the state of the device are invalidated and so are the agents'
  beliefs based on these observations. Other observations, however, may still be valid and

---

[1] Strictly speaking, information gathering may involve *active sensing* in which the sensors may be relocated or
repositioned (e.g., relocation of a video camera or a robot). In such cases, the state of domain may be changed.

were costly to obtain in the first place. How to make use of still valid observations while discarding invalid ones in updating the agents' beliefs is an open issue.

10.3 What If Verification Fails?

In Chapter 9, we presented a set of algorithms for verifying the integrity of integrated multiagent system against the technical requirements of an MSBN. After the verifications succeed, agents will go on to perform compilation and inference operations. What should agents do if some verifications fail? We briefly consider such failures.

- Failure in hypertree verification: The subdomain partition (Section 9.3) among a set of agents may not admit a hypertree organization. In that case, the interface graph that the system integrator created based on its knowledge of public variables is not chordal (Theorem 9.1). The integrator may advise the agents to modify their subdomains by enlarging or reducing public variables. There are usually many possible modifications yielding subdomain partitions that admit hypertree organizations. Some schemes require modifications of subdomains of more agents than other schemes. Some schemes require more significant subdomain modifications for each agent involved than other schemes. Some schemes compromise agent privacy more than other schemes. How the integrator can generate proposals that minimize the subdomain revision and compromise of agent privacy is yet to be explored.
- Failure in acyclicity verification: Once the hypertree organization is verified, the hypertree DAG union may still be acyclic (Section 9.5). The algorithms presented can easily be extended to indicate explicitly which agents are causing the acyclicity. How should these agents cooperate to resolve the problem? How can they modify their local structure without compromising the integrity of their local knowledge representation? How can they minimize the modification to their local structures in trying to satisfy the global acyclicity constraint?
- Failure in d-sepset verification: An agent interface in an acyclic hypertree DAG union may not be a d-sepset (Section 9.6) because some public nodes in the interface are not d-sepnodes. To convert a non-d-sepnode into a d-sepnode, some of its parent nodes may either be removed or become public. Again, the potential compromise of integrity of knowledge representation at each agent involved and the degree of compromise of agent privacy should be considered and minimized.

10.4 Dynamic Formation of MSBNs

Large and complex domains are often open-ended. That is, the set of domain variables grows and shrinks from time to time. To cope with such open-endedness, multiagent systems respond by a changing agent population. As the problem domain grows in size and complexity, more agents with proper knowledge are deployed or activated. When their special knowledge is no longer needed owing to changes in the problem domain, some agents are deactivated. For example, the installation of a new appliance in a smart house requires one or more agents to be deployed with knowledge about the effective utilization of the appliance in the home environment. When the appliance is replaced with a new model, the associated agents must be deactivated and a new set of agents added.

Another need that accounts for the open-endedness of a multiagent system is that of being able to zoom in and zoom out at different abstraction levels. The issue under the single-agent paradigm was studied by Srinivas [69], Kollar and Pfeffer [35] and Wong and Butz [78]. For example, the functionality of a component may be represented at the component level, which is usually sufficient. However, it may sometimes be necessary to examine the component at its device level. The agent knowledgeable at the component level may not have knowledge at the device level, thus necessitating activation and consultation of a proper new agent. In summary, change in agent population arises when a change in domain dimensions or abstraction levels occurs. How can such dynamic formation and reformation be realized without human intervention or with the minimum human intervention in the context of an MSBN-based multiagent system?

- When the need for a new agent arises, how should it be integrated into the existing hypertree organization? It may be possible for the new agent to be added to the hypertree as a terminal, in which case no adjustment for the adjacency among existing agents is needed except the agent adjacent to the new agent. But what if this is not possible? How should the hypertree be restructured in such cases?
- A related issue is the effects of hypertree restructuring on the agents' beliefs. In the case of a new terminal agent, how should the beliefs of the new agent as well as those of existing agents be updated according to the knowledge and observations in the existing system and the knowledge embedded in the new agent? If the hypertree must be restructured significantly, how can belief updating be accomplished?

10.5 Knowledge Adaptation and Learning

Besides the more dramatic changes in problem domain dimensions and abstraction levels discussed in Section 10.4, a problem domain may evolve slowly over time. That is, the space of a given variable may change over time. Some values in the original space may become impossible, and others may need to be introduced. For example, the air conditioning switch will never be on in the winter but may be turned on and off during the summer. The prior distribution of variables and the dependence strength among variables may gradually change over time. As a mechanical component in a piece of equipment wears out, the likelihood of a breakdown increases gradually. Just as learning is important for a single agent to evolve its knowledge of its problem domain, so is it for a multiagent system. It appears that techniques developed for sequential learning in graphical models under the single-agent paradigm, such as those of Spiegelhalter and Lauritzen [68], can be applied directly. Careful theoretical and empirical studies are needed before this is confirmed.

In addition to the model adaptation just described, other aspects of operations an MSBN-based multiagent system's can be made more effective through learning. For example, regional communication (Section 8.9) is less expensive than full-scale communication. However, given an agent that needs to communicate regionally to resolve the uncertainty on a particular variable or a set of variables, which nearby agents should be included in the regional communication must be based on the dependence relations between the target variables and variables in the other agents. Note that we have insisted that the internal details of each agent be kept private. Hence, a centralized sensitivity analysis is out of the question. By learning from past communications experience using different regions, an agent can improve its estimation of the proper region of communication given a set of target variables. How to control and vary communications regions, how to represent past experience regarding different regions concisely, and how to use learning to improve the efficiency of future communications are yet to be explored.

10.6 Negotiation over Agent Interfaces

The cardinality of an agent interface (the d-sepset) in an MSBN may be perceived as the ultimate factor determining the efficiency of communication. Such a perception may originate from the observation that the cardinality of a separator in the JT representation of a BN indeed determines the volume of a concise message passed over the separator and directly affects the inference efficiency. However, this is only partly true for inference in a multiagent MSBN owing to the linkage tree representation for each d-sepset.

Figure 10.3 shows the local JTs of two agents in a trivial MSBN and their interface. The linkage tree L has two linkages. Suppose that the space of each

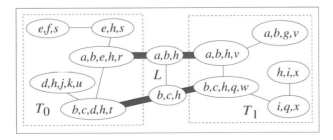

Figure 10.3: A trivial linkage tree.

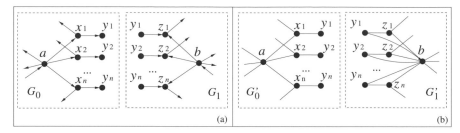

Figure 10.4: (a) Two subnets in an MSBN. (b) The local moral graphs.

variable has the cardinality of 4. The raw-space complexity of L is then $4^3 + 4^3 = 128$, whereas the cardinality of the full-state space of the d-sepset $\{a, b, c, h\}$ is $4^4 = 256$. Because the size of an e-message is determined by the raw-space complexity of the linkage tree, this example illustrates that the raw-space complexity of the linkage tree is a more direct factor for communications efficiency in an MSBN. Furthermore, the following example shows that it is possible to increase the cardinality of a d-sepset while decreasing the raw space complexity of the linkage tree.

Figure 10.4(a) partially shows two adjacent subnets in an MSBN. The d-sepset is $\{y_1, \ldots, y_n\}$. Assume that there are no other paths between the nodes explicitly illustrated in (a) except those that are shown. After moralization, the local moral graphs are those in (b). During cooperative triangulation (Section 7.6), the agent A_0 needs to eliminate $a, x_1, \ldots, x_n$ before y_i ($i = 1, \ldots, n$). As a consequence, the d-sepset will be completed. The resultant linkage tree is trivial and has a raw-space complexity $O(2^n)$.

Suppose that the d-sepset is enlarged by making variables a and b public. The new subnet G_0 is shown in Figure 10.5(a) and its local moral graph in (b). During

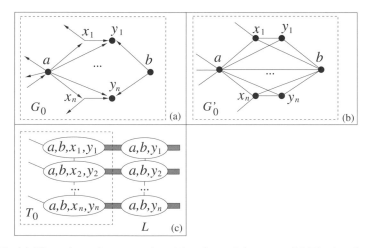

Figure 10.5: (a) The subnet for agent A_0 with enlarged d-sepset. (b) The local moral graph of A_0. (c) The local junction tree of A_0 and its linkage tree with A_1.

cooperative triangulation, A_0 can eliminate $x_1, \ldots, x_n$ before $a, b, y_1, \ldots, y_n$. If it eliminates in the order $(x_1, \ldots, x_n, y_1, \ldots, y_n, a, b)$, no fill-in is added. The resultant partial local JT for A_0 is shown to the left of (c), and the corresponding linkage tree is shown to the right. The linkage tree now has reduced the raw-space complexity from $O(2^n)$ to $O(2^3 n)$. A similar reduction can be obtained in A_1.

In general, such reductions can be obtained when the new public variables help partition each subdomain involved into conditionally independent parts. For example, every two variables in the set $x_1, \ldots, x_n$ are rendered conditionally independent by the new d-sepset.

Note that the modification does not require assessment of a new set of probability distributions. For example, node y_1 is assigned $P(y_1|x_1)$ in the subnet G_0 in Figure 10.4(a). With the new subnet G_0 in Figure 10.5(a), there is no need to assess $P(y_1|x_1, a, b)$ and reassign y_1 this distribution. The reason is that the remodeling does not alter any dependence relations among variables. It simply exploits some existing independence in the total universe. As a consequence, the JPD of the new MSBN is identical to the previous one, and no node needs to change its previously assigned distribution in a nontrivial way.

This phenomenon gives rise to a promising opportunity. That is, agents may explore this opportunity to reduce the raw-space complexity of their linkage trees and ultimately improve the efficiency of their communications. Because the internal structure of each agent is private, identification of a potentially useful enlargement of public variables can only be achieved through distributed search and cooperation among agents. Inasmuch as some previously private variables must be made public, negotiation among agents may be needed as well.

10.7 Relaxing Hypertree Organization

In Section 6.3, we introduced Basic Assumption 6.3 stating that a simpler agent organization (as a subgraph of the communications graph) is preferred. It asserts that even though the communications graph with degenerate cycles can support belief updating by message passing, a hypertree organization is preferred because it is simpler and more efficient to process. The hypertree restriction is sometimes perceived as "undesirable" because agents are prevented from communicating freely with each other. Flexibility is the main reason to desire its relaxation.

Using the graph-theoretical equivalence between cluster graphs and communications graphs (Chapter 6), we can extend the results on degenerate and nondegenerate cycles in Chapter 3 to multiagent systems. In Chapters 3 and 6, we have shown that when nondegenerate cycles exist in the communications graph, exact multiagent belief updating by concise message passing is impossible. Hence, only degenerate cycles offer an opportunity for agents to communicate more freely. A degenerate cycle in a communications graph is signified by the presence of an agent interface contained in every other agent interface on the cycle. Hence, if a set of agents is located on a strong degenerate cycle in their communications graph, they can pass e-messages to each other in arbitrary sequences. For weak degenerate cycles, the flexibility is more limited. Consider a weak degenerate cycle $\rho = \langle Q_0, Q_1, \ldots, Q_{K-1}, Q_0 \rangle$, where $K \geq 3$ and the separator $S = Q_0 \cap Q_1$ is contained in each other separator. Two paths exist between any two clusters: one goes through S and the other does not. Because each separator in the other path is a superset of S, the information capacity of the path through S is inferior. The exact degree of flexibility that weak degenerate cycles can provide needs to be further explored.

In summary, exploration of degenerate cycles in communications graphs offers a promising possibility for relaxing the hypertree constraint when greater flexibility in communications is important. Existing degenerate cycles can be utilized to its full capacity in providing such flexibility and agent interfaces can be enlarged to create new strong degenerate cycles. The trade-off involves the discloser of some private variables and increased message traffic.

10.8 Model Approximation

Multiply sectioned Bayesian networks provide a framework in which a set of agents populating a large and complex uncertain total universe can reason about it by concise message passing. The inference computation is efficient when the dependence relations in the total universe are sparse. That is, individual subnets are not densely connected and individual agent interfaces have low cardinality and can be concisely represented by linkage trees. On the other hand, if subnets are densely connected, agent interfaces are large when determined naturally, or agent interfaces do not admit concise representation, the communications and inferences of an MSBN-based multiagent system can be computationally expensive.

One alternative that can be explored within the MSBN framework is *model approximation*, which entails simplifying some subnets. The most obvious simplification is the removal of arcs corresponding to weak dependence relations. The positive consequences are not only sparser subnets but more concise agent interface representations (lower raw-space complexity of linkage trees). Because the weakest arcs are continuously removed, it is expected that the efficiency of inference and communication improves gradually whereas the accuracy of inference degrades gradually.

It is not clear what approximation can be performed when the cardinality of an agent interface is too high. One possibility is to reduce the number of public variables. Issues arise in choosing the variables to eliminate from the agent interface and deciding how to turn them into private variables. Another possibility is to explore the technique outlined in Section 10.6. Instead of trying to reduce the agent interface, it can be enlarged so that the raw-space complexity of the corresponding linkage tree is reduced.

10.9 Mixed Models

Throughout the book, we have concentrated on the total universe consisting of only discrete variables. Many problem domains involve continuous variables – that is, variables whose spaces are continuous ranges. Although a continuous variable can always be discretized, discretization into a low-cardinality space introduces errors, whereas discretization into a high-cardinality space increases the computational complexity. Hence, there are cases in which direct representation of, and inference with, continuous variables are desirable. One approach under the single-agent paradigm that has been well studied is the use of *multivariate Gaussian distribution* (Lauritzen [36], Cowell et al. [9]). Both discrete and continuous variables can be present in a single graphical model. Extending the MSBN framework following this approach is yet to be explored.

In Chapter 6, we posited Basic Assumption 6.4, which states that a DAG structuring of each agent's knowledge is preferred. In Chapter 7, we compiled each subnet into a junction tree representation. Although a causal structuring of probabilistic knowledge is commonly preferred (see Basic Assumption 6.4), the possibility of performing local inference and communication using the junction tree representation means that causal structuring is not necessary. Undirected graphical models such as *Markov networks* (Pearl [52]), and hybrid graphical models such as *chain graph models* (Lauritzen [36]) have been studied under the single-agent paradigm. It is conceivable that different agents may adopt different graphical models as their original internal representation. For local inference and communication, the local models can be compiled into the junction tree representation, and the algorithms in Chapter 8 can then be applied. The technical details of such a scheme are yet to be worked out.

10.10 Bibliographical Notes

Dynamic Bayesian networks (DBNs) were proposed by Dean and Kanazawa [15]. Russel and Norvig [60] gave an intuitive introduction of reasoning in DBNs. Kjaerulff [34] and Xiang [86] have proposed exact methods for inference in DBNs, and Forbes et al. [17] have presented an approximate method. The opportunity for reducing raw-space complexity for agent interface was analyzed by Xiang et al. [93].

Bibliography

[1] O. Bangso and P.H. Wuillemin. Top-down construction and repetitive structures representation in Bayesian networks. In *Proc. 13th Inter. Florida Artificial Intelligence Research Society Conf.* AAAI Press: Menlo Park, CA, 2000.

[2] C. Beeri, R. Fagin, D. Maier, and M. Yannakakis. On the desirability of acyclic database schemes. *J. ACM*, 30(3):479–513, 1983.

[3] D.S. Bernstein, S. Zilberstein, and N. Immerman. The complexity of decentralized control of Markov decision processes. In *Proc. 16th Conf. on Uncertainty in Artificial Intelligence*, Stanford, 2000.

[4] M. Boman, P. Davidsson, and H.L. Younes. Artificial decision making under uncertainty in intelligent buildings. In K.B. Laskey and H. Prade, editors, *Proc. 15th Conf. on Uncertainty in Artificial Intelligence*, pages 65–70, Stockholm, 1999.

[5] A.H. Bond and L. Gasser, editors. *Readings in Distributed Artificial Intelligence*. Morgan Kaufmann: San Mateo, CA, 1988.

[6] E. Castillo, J. Gutierrez, and A. Hadi. *Expert Systems and Probabilistic Network Models*. Springer-Verlag: New York, 1997.

[7] P. Cheeseman. An inquiry into computer understanding. *Computational Intelligence*, 4(1):58–66, 1988.

[8] G.F. Cooper. A method for using belief networks as influence diagrams. In R.D. Shachter, T.S. Levitt, L.N. Kanal, and J.F. Lemmer, editors, *Proc. 4th Workshop on Uncertainty in Artificial Intelligence*, pages 55–63, 1988.

[9] R.G. Cowell, A.P. Dawid, S.L. Lauritzen, and D.J. Spiegelhalter. *Probabilistic Networks and Expert Systems*. Springer-Verlag: New York, 1999.

[10] P. Dagum and M. Luby. Approximating probabilistic inference in Bayesian belief networks is NP-hard. *Artificial Intelligence*, 60(1):141–153, 1993.

[11] R. Davis. Diagnostic reasoning based on structure and behavior. *Artificial Intelligence*, 24:347–410, 1984.

[12] A.P. Dawid and S.L. Lauritzen. Hyper Markov laws in the statistical analysis of decomposable graphical models. *Annals of Statistics*, 21(3):1272–1317, 1993.

[13] J. de Kleer and B.C. Williams. Diagnosing multiple faults. *Artificial Intelligence*, (32):97–130, 1987.

[14] T. Dean, J. Allen, and Y. Aloimonos. *Artificial Intelligence: Theory and Practice*. Benjamin/Cummings: Boston, 1995.

[15] T.L. Dean and K. Kanazawa. A model for reasoning about persistence and causation. *Computational Intelligence*, 5:142–150, 1989.

[16] M.J. Druzdzel and R.R. Flynn. Decision support systems. In A. Kent, editor, *Encyclopedia of Library and Information Science*. Marcel Dekker: New York, 2000.

[17] J. Forbes, T. Huang, K. Kanazawa, and S. Russell. The batmobile: Towards a Bayesian automated taxi. In *Proc. 14th Inter. Joint Conf. on Artificial Intelligence*, pages 1878–1885, Montreal, 1995.

[18] R. Fung and B.D. Favero. Backward simulation in Bayesian networks. In R.L. de Mantaras and D. Poole, editors, *Proc. 10th Conf. on Uncertainty in Artificial Intelligence*, pages 227–234, 1994.

[19] S. Geman and D. Geman. Stochastic relaxation, Gibbs distributions, and the Bayesian restoration of images. *IEEE Trans. on Pattern Analysis and Machine Intelligence*, 6(6):721–741, 1984.

[20] M.R. Genesereth. The use of design descriptions in automated diagnosis. *Artificial Intelligence*, 24:411–436, 1984.

[21] M.C. Golumbic. *Algorithmic Graph Theory and Perfect Graphs*. Academic Press: San Francisco, 1980.

[22] R.P. Grimaldi. *Discrete and Combinatorial Mathematics: An Applied Introduction*. Addison–Wesley: Reading, MA, 1989.

[23] M.H. Hassoun. *Fundamentials of Artificial Neural Networks*. MIT Press: Cambridge, MA, 1995.

[24] M. Henrion. Propagating uncertainty in Bayesian networks by probabilistic logic sampling. In J.F. Lemmer and L.N. Kanal, editors, *Uncertainty in Artificial Intelligence 2*, pages 149–163. Elsevier Science Publishers: Amsterdam, 1988.

[25] J. Holland. Earthquake prediction – Chinese style. Alaska Science Forum, May 1976.

[26] R.A. Howard and J.E. Matheson. Influence diagrams. In R.A. Howard and J.E. Matheson, editors, *Readings on the Principles and Applications of Decision Analysis*, pages 721–762. Menlo Park, CA: Strategic Decisions Group, 1984.

[27] C.S. Jensen, A. Kong, and U. Kjarulff. Blocking–Gibbs sampling in very large probabilistic expert systems. *Inter. J. Human–Computer Studies*, 42:647–666, 1995.

[28] F.V. Jensen. Junction tree and decomposable hypergraphs. Technical report, JUDEX, Aalborg, Denmark, February 1988.

[29] F.V. Jensen. *An Introduction to Bayesian Networks*. UCL Press: London, 1996.

[30] F.V. Jensen, S.L. Lauritzen, and K.G. Olesen. Bayesian updating in causal probabilistic networks by local computations. *Computational Statistics Quarterly*, 4:269–282, 1990.

[31] R.M. Jones, J.E. Laird, P.E. Nielsen, K.J. Coulter, P. Kenny, and F.V. Koss. Automated intelligent pilots for combat flight simulation. *AI Magazine*, 20(1):27–41, 1999.

[32] D. Kahneman, P. Slovic, and A. Tversky, editors. *Judgment under uncertainty: Heuristics and biases*. Cambridge University Press: New York, 1982.

[33] J.H. Kim and J. Pearl. A computational model for combined causal and diagnostic reasoning in inference systems. In *Proc. 8th Inter. Joint Conf. on Artificial Intelligence*, pages 190–193, 1983.

[34] U. Kjaerulff. A computational scheme for reasoning in dynamic probabilistic networks. In D. Dubois, M.P. Wellman, B. D'Ambrosio, and P. Smets, editors, *Proc. 8th Conf. on Uncertainty in Artificial Intelligence*, pages 121–129, Stanford, CA, 1992.

[35] D. Koller and A. Pfeffer. Object-oriented Bayesian networks. In D. Geiger and P.P. Shenoy, editors, *Proc. 13th Conf. on Uncertainty in Artificial Intelligence*, pages 302–313, Providence, Rhode Island, 1997.

[36] S.L. Lauritzen. *Graphical Models*. Oxford Science: New York, 1996.

[37] S.L. Lauritzen and D.J. Spiegelhalter. Local computation with probabilities on graphical structures and their application to expert systems. *J. Royal Statistical Society, Series B*, 50:157–244, 1988.

[38] V.R. Lesser and L.D. Erman. Distributed interpretation: A model and experiment. *IEEE Trans. on Computers*, C-29(12):1144–1163, 1980.

[39] A.L. Madsen and F.V. Jensen. Lazy propagation in junction trees. In *Proc. 14th Conf. on Uncertainty in Artificial Intelligence*, 1998.

[40] D. Maier. *The Theory of Relational Databases*. Computer Science Press, 1983.

[41] G.E. Monahan. A survey of partially observable Markov decision processes: Theory, models, and algorithms. *Management Science*, (1):1–16, 1982.

[42] K.P. Murphy, Y. Weiss, and M.I. Jordan. Loopy belief propagation for approximate inference: An empirical study. In K.B. Laskey and H. Prade, editors, *Proc. 15th Conf. on Uncertainty in Artificial Intelligence*, pages 467–475, Stockholm, 1999.

[43] R.E. Neapolitan. *Probabilistic Reasoning in Expert Systems*. John Wiley and Sons: New York, 1990.

[44] N.J. Nilsson. *Artificial Intelligence: A New Synthesis*. Morgan Kaufmann: San Francisco, 1998.

[45] R.M. Oliver and J.Q. Smith, editors. *Influence Diagrams, Belief Nets and Decision Analysis*. John Wiley: New York, 1990.

[46] L.E. Ortiz and L.P. Kaelbling. Adaptive importance sampling for estimation in structured domains. In C. Boutilier and M. Goldszmidt, editors, *Proc. 16th Conf. on Uncertainty in Artificial Intelligence*, pages 446–454, 2000.

[47] H.V.D. Parunak. Industrial and practical applications of DAI. In G. Weiss, editor, *Multiagent Systems: A Modern Approach to Distributed Artificial Intelligence*, pages 377–421. MIT Press: Cambridge, MA, 1999.

[48] J. Pearl. Reverend Bayes on inference engines: A distributed hierarchical approach. In *Proc. National Conference on Artificial Intelligence*, pages 133–136, 1982.

[49] J. Pearl. A constraint-propagation approach to probabilistic reasoning. In L.N. Kanal and J.F. Lemmer, editors, *Uncertainty in Artificial Intelligence*, Elsevier Science Publishers: Amsterdam, pages 357–370, 1986.

[50] J. Pearl. Fusion, propagation, and structuring in belief networks. *Artificial Intelligence*, 29(3):241–288, 1986.

[51] J. Pearl. Evidential reasoning using stochastic simulation of causal models. *Artificial Intelligence*, 32(2):245–257, 1987.

[52] J. Pearl. *Probabilistic Reasoning in Intelligent Systems: Networks of Plausible Inference*. Morgan Kaufmann: San Francisco, 1988.

[53] A. Pfeffer, D. Koller, B. Milch, and K.T. Takusagawa. SPOOK: A system for probabilistic object-oriented knowledge representation. In K.B. Laskey and H. Prade, editors, *Proc. 15th Conf. on Uncertainty in Artificial Intelligence*, pages 541–550, Stockholm, 1999.

[54] D. Poole. Probabilistic horn abduction and Bayesian networks. *Artificial Intelligence*, 64(1):81–129, 1993.

[55] D. Poole, A. Mackworth, and R. Goebel. *Computational Intelligence: A Logical Approach*. Oxford University Press: New York, 1998.

[56] M. Pradhan, G. Provan, B. Middleton, and M. Henrion. Knowledge engineering for large belief networks. In *Proc. 10th Conf. Uncertainty in Artificial Intelligence*, pages 484–490, Seattle, 1994.

[57] M.L. Puterman. *Markov Decision Processes*. John Wiley: New York, 1994.

[58] D.J. Rose, R.E. Tarjan, and G.S. Lueker. Algorithmic aspects of vertex elimination on graphs. *SIAM J. Computing*, 5:266–283, 1976.

[59] J.S. Rosenschein and G. Zlotkin. *Rules of Encounter.* MIT Press: Cambridge, MA, 1994.

[60] S. Russell and P. Norvig. *Artificial Intelligence: A Modern Approach.* Prentice-Hall: Englewood Cliffs, NJ, 1995.

[61] R.D. Shachter and M.A. Poet. Simulation approaches to general probabilistic inference on belief networks. In *Proc. 5th Workshop on Uncertainty in Artificial Intelligence*, pages 311–318, Windsor, ON, 1989.

[62] G. Shafer. *Probabilistic Expert Systems.* Society for Industrial and Applied Mathematics, Philadelphia, 1996.

[63] G. Shafer and J. Pearl, editors. *Readings in Uncertain Reasoning.* Morgan Kaufmann: San Mateo, CA, 1990.

[64] G. Shafer, P.P. Shenoy, and K. Mellouli. Propagating belief functions in qualitative Markov trees. *Inter. J. of Approximate Reasoning*, (1):349–440, 1987.

[65] A. Silberschatz and P.B. Galvin. *Operating System Concepts.* Addison-Wesley: Reading, MA, 1998.

[66] R.D. Smallwood and E.J. Sondik. The optimal control of partially observable Markov processes over a finite horizon. *Operation Research*, 21:1071–1088, 1973.

[67] D.J. Spiegelhalter. Probabilistic reasoning in predictive expert systems. In L.N. Kanal and J.F. Lemmer, editors, *Uncertainty in Artificial Intelligence*, pages 47–68. Elsevier Science Publishers: Amsterdam, 1986.

[68] D.J. Spiegelhalter and S.L. Lauritzen. Sequential updating of conditional probabilities on directed graphical structures. *Networks*, 20:579–605, 1990.

[69] S. Srinivas. A probabilistic approach to hierarchical model-based diagnosis. In *Proc. 10th Conf. Uncertainty in Artificial Intelligence*, pages 538–545, Seattle, 1994.

[70] M. Studeny. Conditional independence relations have no finite complete characterization. In S. Kubik and J.A. Visek, editors, *Information Theory, Statistical Decision Functions and Random Processes*, pages 377–396. Kluwer: Dordrecht, 1992.

[71] K.P. Sycara. Multiagent systems. *AI Magazine*, (2):79–92, 1998.

[72] P. Szolovits and S.G. Pauker. Categorical and probabilistic reasoning in medical diagnosis. *Artificial Intelligence*, 11:115–144, 1978.

[73] A.S. Tanenbaum. *Distributed Operating Systems.* Prentice-Hall: Englewood Cliffs, NJ, 1995.

[74] R.E. Tarjan and M. Yannakakis. Simple linear time algorithms to test chordality of graphs, test acyclicity of hypergraphs, and selectively reduce acyclic hypergraphs. *SIAM J. of Computing*, 13(3):566–579, 1984.

[75] R.L. Teach and E.H. Shortliffe. An analysis of physician attitudes regarding computer-based clinical consultation systems. *Computers and Biomedical Research*, (14):542–558, 1981.

[76] L.C. van der Gaag and J.J.Ch. Meyer. Informational independence: Models and normal forms. *Inter. J. of Intelligent Systems*, 13:83–109, 1998.

[77] G. Weiss, editor. *Multiagent Systems: A Modern Approach to Distributed Artificial Intelligence.* MIT Press: Cambridge, MA, 1999.

[78] S.K.M. Wong and C.J. Butz. Contextual weak independence in Bayesian networks. In K.B. Laskey and H. Prade, editors, *Proc. 15th Conf. on Uncertainty in Artificial Intelligence*, pages 670–679. Morgan Kaufmann: San Francisco, 1999.

[79] M. Wooldridge and N.R. Jennings. Intelligent agents: Theory and practice. *Knowledge Engineering Review*, 10(2):115–152, 1995.

[80] Y. Xiang. *Multiply sectioned Bayesian networks for large knowledge-based systems and their applications in neuromuscular diagnosis.* Ph.D. thesis, University of British Columbia, 1992.

[81] Y. Xiang. Distributed multi-agent probabilistic reasoning with Bayesian networks. In Z.W. Ras and M. Zemankova, editors, *Methodologies for Intelligent Systems*, pages 285–294. Springer-Verlag: Berlin, 1994.

[82] Y. Xiang. Distributed scheduling of multiagent systems. In *Proc. 1st Inter. Conf. on Multi-agent Systems*, pages 390–397, San Francisco, 1995.

[83] Y. Xiang. Optimization of inter-subnet belief updating in multiply sectioned Bayesian networks. In *Proc. 11th Conf. on Uncertainty in Artificial Intelligence*, pages 565–573, Montreal, 1995.

[84] Y. Xiang. A probabilistic framework for cooperative multi-agent distributed interpretation and optimization of communication. *Artificial Intelligence*, 87(1–2):295–342, 1996.

[85] Y. Xiang. A characterization of single-link search in learning belief networks. In P. Compton, H. Motoda, R. Mizoguchi, and H. Liu, editors, *Proc. Pacific Rim Knowledge Acquisition Workshop*, pages 218–233, Singapore, 1998.

[86] Y. Xiang. Temporally invariant junction tree for inference in dynamic Bayesian network. In R.E. Mercer and E. Neufeld, editors, *Advances in Artificial Intelligence*, pages 363–377. Springer-Verlag: Berlin, 1998.

[87] Y. Xiang. Verification of DAG structures in cooperative belief network-based multi-agent systems. *Networks*, 31:183–191, 1998.

[88] Y. Xiang. Belief updating in multiply sectioned Bayesian networks without repeated local propagations. *Inter. J. Approximate Reasoning*, 23:1–21, 2000.

[89] Y. Xiang. Cooperative triangulation in MSBNs without revealing subnet structures. *Networks*, 37(1):53–65, 2001.

[90] Y. Xiang and H. Geng. Distributed monitoring and diagnosis with multiply sectioned Bayesian networks. In *Proc. AAAI Spring symposium on AI in Equipment Service, Maintenance and Support*, pages 18–25, Stanford, 1999.

[91] Y. Xiang and F.V. Jensen. Inference in multiply sectioned Bayesian networks with extended Shafer–Shenoy and lazy propagation. In *Proc. 15th Conf. on Uncertainty in Artificial Intelligence*, pages 680–687, Stockholm, 1999.

[92] Y. Xiang and V. Lesser. Justifying multiply sectioned Bayesian networks. In *Proc. 6th Inter. Conf. on Multi-agent Systems*, pages 349–356, Boston, 2000.

[93] Y. Xiang, K.G. Olesen, and F.V. Jensen. Practical issues in modeling large diagnostic systems with multiply sectioned Bayesian networks. *Inter. J. Pattern Recognition and Artificial Intelligence*, 14(1):59–71, 2000.

[94] Y. Xiang, B. Pant, A. Eisen, M.P. Beddoes, and D. Poole. PAINULIM: A neuromuscular diagnostic aid using multiply sectioned Bayesian networks. In *Proc. ISMM Inter. Conf. on Mini and Microcomputers in Medicine and Healthcare*, pages 64–69, Long Beach, 1991.

[95] Y. Xiang, B. Pant, A. Eisen, M.P. Beddoes, and D. Poole. Multiply sectioned Bayesian networks for neuromuscular diagnosis. *Artificial Intelligence in Medicine*, 5:293–314, 1993.

[96] Y. Xiang, D. Poole, and M.P. Beddoes. Multiply sectioned Bayesian networks and junction forests for large knowledge-based systems. *Computational Intelligence*, 9(2):171–220, 1993.

[97] M. Yannakakis. Computing the minimum fill-in is NP-complete. *SIAM J. of Algebraic and Discrete Methods*, 2(1), 1981.

Index

293